Citizens Band Transceivers: Installation and Troubleshooting

Citizens Band Transceivers: Installation and Troubleshooting

MANNIE HOROWITZ

Reston Publishing Company, Inc., Reston, Virginia
A Prentice-Hall Company

Library of Congress Cataloging in Publication Data

Horowitz, Mannie.
    Citizens band transceivers.

    Includes index.
    1. Citizens band radio—Equipment and supplies.
I. Title.
TK6570.C5H67   621.3845′4   77-8510
ISBN 0-87909-102-9

© 1978 by
Reston Publishing Company, Inc.
A Prentice-Hall Company
Reston, Virginia 22090

All rights reserved. No part of this
book may be reproduced in any way,
or by any means, without permission
in writing from the publisher.

10 9 8 7 6 5 4 3 2 1

Printed in the United States of America.

To my wife Ruth and my daughter Beverly

Contents

Preface, xi

Chapter 1   **Introduction to Citizens Band Radio**   1

Station Classifications and Characteristics, 2
License and Equipment, 4
Important Restrictions, 6
Adding to Your Efficiency, 7

Chapter 2   **Transceiver**   13

The Basic Radio, 13
Transceiver Features, 19
Microphone, 27
Antennas, 31

Chapter 3 Accessories 33

Standing-Wave-Radio/Power/Field-Strength Meters, 34
Power Indicators, 37
Antenna Matcher, 38
Antenna Selector, 39
Two-Transceiver Couplers, 39
Cophaser, 40
Lightning Arrestor, 41
Modulation Meter, 42
Television-Interference Filters, 43
Power Supply, 44
Remote Loudspeaker and Headphone, 46

Chapter 4 Tests and Test Equipment 49

Measuring Voltage and Resistance, 50
Power Supply, 53
Transistor Testers, 55
Tube Tester, 69
Oscilloscope, 70
Signal Tracer, 74
Signal Injector, 75
Signal Generators, 76
Digital Frequency Counter, 79
Dip Meter, 81
Radio-Frequency Wattmeter, 82
Special Instruments for Use in the Field, 85

Chapter 5 Troubleshooting the Transmitter 87

Amplitude-Modulation Transmitter Blocks, 88
Single-Sideband Transmitter Blocks, 119

Chapter 6 Troubleshooting the Receiver 129

Radio-Frequency Amplifier, 130
Local Oscillator, Mixer and Intermediate-Frequency Stages, 134
Detectors, 137

Automatic Noise Limiter, 138
Noise Blanking, 139
Troubleshooting Detectors and Noise-Deleting Circuits, 140
Squelch Circuit, 141
Audio Amplifier Circuits, 144
Aligning the Receiver, 145
Single-Sideband Receivers, 147

Chapter 7 **Troubleshooting the Complete Transceiver**   151

Dual Conversion Transceiver, 151
Practical Troubleshooting Procedures, 155
Public-Address Operation, 166
An Innovation, 166

Chapter 8 **Choosing a Microphone**   169

Troubleshooting Microphones, 170
Replacement Microphones, 173
Amplified Microphones, 177
Compressor Amplifiers, 180
Peak Redistribution Modulation, 181

Chapter 9 **Antennas**   185

Antenna Characteristics, 186
Supporting Structures, 191
Rotators, 192
Coaxial Cable, 192
Base Station Antennas, 195
Mobile Antennas, 200
Tuning the Antenna, 205

Chapter 10 **Problems Encountered in the Field**   209

Installing the CB Radio, 210
Antennas, 215
Power Supply Problems, 217

Index, 219

Preface

Citizens band radio had its modest beginnings in 1958. It was a sleeping giant for a number of years. Only the businessman, the wealthy individual, and the avid radio enthusiast were attracted. From a few thousand people using CB transceivers several years ago, CB users are now numbered in the millions. The FCC just cannot keep up with the many requests for licenses.

Millions of 23-channel radios are now in use. Because many CBers broadcast on just a few of these 23-channels, it is very unlikely that they will discard their radios for the newer 40-channel jobs. The old radios are frequently more efficient than the newer models due to the tightening of radiation specifications on all CB radios sold after January 1, 1977. In order to meet FCC requirements, some of the 40-channel radios are designing to deliver less than the allowed maximum of four watts.

Nearly 1,000 different 23-channel CB radios are now on the market and many new 40-channel transceivers are reaching the dealer's shelves every day; therefore, installing and servicing CB rigs is a big business. This book is written to provide necessary information to perform the servicing function most efficiently. It is directed primarily to both the experienced service technician who has a general knowledge of electronic circuits and techniques, as well as to the new technician who is either now receiving on-the-job training or is a recent graduate of a technical school. The CB enthusiast and do-it-yourselfer have not been left out of the picture. Chapters have been included which: introduce CB to the reader, discuss CB radio briefly in not-too-technical terms, detail installation techniques that anyone can perform, and describe the many accessories useful to the CBer and how to use them. The engineer has also not been forgotten—this book will introduce him to the many peculiarities of CB radio which he can apply to his designs.

Following several introductory chapters, a detailed discussion of applicable test instruments is pursued. When specialized equipment such as transistor testers is described, a review presenting basic principles of the devices to be checked in the instrument is included. This review has a two-fold purpose. First, enough component information is provided so that the devices encountered in CB radio circuits are not foreign to the new technician and enthusiast. Second, device and associated test instrument information is provided in order that the technician and hobbyist will make a wise choice of equipment that best suits their needs. Some applications of the various instruments are elaborated upon in Chapter 4.

Subsequent chapters deal with the various sections of the transceiver proper. An artificial division is made between the transmitter and receiver sections. The CB radio is then treated as a complete item when a practical circuit is dissected and troubleshooting procedures are recommended. Some emphasis is placed on the phase-locked loop and single-sideband radios due to their increasing importance in the modern 40-channel rigs.

Antennas and microphones are *not* treated as stepchildren of the setup. They are vital factors in an efficient installation. Practical considerations include detailed discussion of the important aspects of the devices which do and do not affect the efficiency of the rig. I point out which concepts are fact and which are fantasy.

After all recommended installation procedures are meticulously pursued and the rig has been completely assembled, Murphy's Law comes into play. Briefly, the law states that if anything can go wrong, it will! The final chapter of the book is devoted to overcome many of the unexpected problems encountered in various setups.

Preface

A good deal of the information, photographs and schematics used in this book were supplied by various transceiver and accessory manufacturers. I therefore extend my sincere thanks to them for their unceasing cooperation in providing me with the necessary material.

MANNIE HOROWITZ

1

Introduction to Citizens Band Radio

In 1947, commercial radio started its shift from the serial and variety show format to the all music and talk show fair that we cherish today. Television was in its infancy and Uncle Miltie was selling gasoline from Maine to Mexico. Not foreseeing the vast wasteland available within the television channels, the Federal Communications Commission (FCC) initiated citizens radio service in a band of UHF frequencies ranging from 460 to 470 megahertz (MHz). This was to allow individuals to communicate with each other by radio as well as to aid business in vehicle dispatching. The advantages of Citizens Band (CB) radio were limitless, but the price was exorbitant. Equipment was extremely expensive, and the distance or range covered by the communications was limited because of the frequencies used. These were referred to as Class A and Class B CB services. Class A was permitted to use transmitters capable of providing much more power to the radiating antenna than Class B.

    The CB service popular today is referred to as Class D Citizens Band Radio Service. A portion of that CB band was introduced in 1958.

There were 23 channels occupying frequencies in the 26.96- to 27.23-MHz band. This was the size of the band until January 1, 1977. After sporadic signs of acceptance, this 23-channel service had little life until the early 1960s, when the Highway Emergency Location Plan (HELP) was developed. All trucks licensed for interstate commerce were required to have CB units in their cabs and to use a CB emergency channel, channel 9 (27.065 MHz), to transmit information concerning hazards on the road, motorists in trouble, and the like. These messages were to be monitored by public authorities capable of furnishing the assistance required. Although the public safety factors involving the truckers failed, it did make them CB conscious. Many transceivers were installed in trucks and were frequently used by the drivers to advise each other of the location of speed traps and weighing stations.

The real use of CB radio was finally realized in the early 1970s. Many motorists found it a useful tool in helping them to avoid heavy traffic on the highway and radar traps monitoring a vehicle's speed to assure that it is not exceeding the legal 55-mph limit. Although the latter use of CB cannot be commended, the location of jam ups on the highway is invaluable information to the law-abiding motorist.

Self-preservation is another important factor that gives the CB business validity. Unless you bury your head in the sand, you know that muggings and other crimes on the streets are grim realities. In some areas, citizens have banded together to alleviate this crime problem by patrolling their neighborhood streets in CB-radio-equipped cars. Should they see any crimes or irregularities, they immediately contact their base station, which relays the information to the police for action.

And, of course, we should not forget the individual who just cannot wait to get home before talking to his wife. Using his CB radio, he may tell her when he should be expected home and at what time to have dinner ready (the male chauvinist). Or he could advise her that he is bringing company home for dinner so that she has sufficient time to sweep the dirt under the rug while adding more water to the soup. Communications can be established between cars, from the car to the home, from your boat to a base station, or from a tractor to the dining hall. The applications are virtually limitless.

Station Classifications and Characteristics

At one time, four classes of CB stations were licensed by the FCC. Class A and Class B stations occupied the band from 460 to 470 MHz. Class A operation was permitted with output power up to 50 watts (W), but Class

Station Classifications and Characteristics

B stations were not allowed anywhere near this signal strength. Because of the limited range at these frequencies, Class B operations were discontinued. Class B radio has not been permitted since November 1971.

Class C CB radio occupies frequencies from 72 to 76 MHz, as well as several frequencies in the standard band for Class D operation from 26.96 to 27.41 MHz. On the upper band, a Class C license entitles the operator to use radio controls for various wireless models for hobby purposes only. On the band shared with Class D users, Class C licensees may operate various devices by remote control. The function of such remote-control activity may be either for self-amusement or to attract attention for advertising and other purposes.

As of this writing, there is a proposal for Class E CB service in the 216- to 218-MHz or 220- to 225-MHz band. This could alleviate some of the overcrowding on other bands. Although the FCC indicated that it would make this band available to the CBer in 1975, nothing has happened as yet.

In this book, we are interested primarily in Class D radiotelephone operation by the United States citizen. As of January 1, 1977, 40 channels are on the CB band, as indicated in Table 1-1. Only 23 of these channels were permitted prior to that date. Transceivers sold before January 1, 1977, accommodate only 23 channels. Originally, the FCC decreed that these CB radios could not legally be modified to accommodate all 40 channels, but then changed its mind. Modifications by manufacturers are permitted, but outboard converters may not be used for this purpose.

All channels but channel 9 can be used for communication between different stations. Channel 9 is reserved solely for emergencies, and may be used only to protect an individual's life or property or when assistance is required by a motorist in trouble. The latter case is not limited only to reporting an accident. Channel 9 may also be used to help a motorist who is out of gas on a highway or in any other similarly untenable position.

The use of channel 9 as an emergency outlet is prescribed by the FCC. Each local area may restrict additional channels to specific uses. When passing through a locality, you should know their specific rules so as not to infringe on anyone's right to use the CB channels as prescribed in that particular area.

Unofficially, maritime enthusiasts have usurped all rights to channel 13. Operators of vehicles in transit feel that channel 19 is exclusively in their domain, and have even extended their claims to cover channels 10 and 12. Channels 16 and 18 are used primarily by single sideband enthusiasts (see Chapter 2). They work the lower sideband of channel 16 and the upper sideband of channel 18. Officially, upper and lower single sideband broadcasting is permitted on all 40 channels.

Table 1-1   Forty Channels for Use by CB Class D Licensees

Channel	Frequency (MHz)	Channel	Frequency (MHz)
1	26.965	21	27.215
2	26.975	22	27.225
3	26.985	23	27.255
4	27.005	24	27.235
5	27.015	25	27.245
6	27.025	26	27.265
7	27.035	27	27.275
8	27.055	28	27.285
9	27.065	29	27.295
10	27.075	30	27.305
11	27.085	31	27.315
12	27.105	32	27.325
13	27.115	33	27.335
14	27.125	34	27.345
15	27.135	35	27.355
16	27.155	36	27.365
17	27.165	37	27.275
18	27.175	38	27.385
19	27.185	39	27.395
20	27.205	40	27.405

Until January 1, 1977, channel 11 was set aside for CBers to contact each other and then switch to another channel to converse. Now there is no official "calling channel." However, any one channel may be agreed upon to unofficially serve this function in a particular area.

License and Equipment

License application forms are readily available from a local FCC office or the main office (Washington, D.C. 20554). These forms are also included with the instruction manual for your new transceiver. The license application form is simple to fill out. Complete it in every respect and mail it to the FCC along with the $4 fee. This license fee covers a five-year period. You can renew it for an additional period of time by simply filing another form 60 days before the expiration date.

It is illegal to use the transceiver (except for receiving purposes) unless you have a license. If you are too impatient to wait for your permanent license to arrive from the FCC, fill out form 555-B. This is a temporary license permitting you to transmit without waiting for your

License and Equipment 5

permanent license. That is, you are permitted to transmit if you have already mailed application form 505 for a permanent license to the FCC. Form 555-B permits you to use your CB rig for 60 days while awaiting the permanent document. It should be delivered within the two-month grace period.

The license should be posted in a conspicuous place if you are broadcasting from a fixed location. Otherwise, obtain form 452-C, a transmitter ID card, from the FCC. This should be attached to each transceiver used in mobile applications, as well as to transceivers installed in fixed locations not readily accessible to the license holder.

Licenses are restricted to citizens of the United States. They cannot be granted to a foreign government or to the representative of a foreign government. A U.S. mailing address must be supplied. Furthermore, the licensee must be 18 years of age or more. However, a Class C license can be secured by anyone over the age of 12.

The licensee is responsible for the transmitter that he uses. He may not make any repairs on the transmitter section unless he holds a first- or second-class radiotelephone license. He is permitted to replace the microphone and antenna, check the transmitter's frequency and modulation level, replace crystals certified as proper by the manufacturer, and align the receiver. Only the repairman with a first- or second-class radiotelephone license is permitted to replace components in the transmitter, install uncertified crystals, alter the transmitter circuit, or adjust its oscillator frequency.

If the date of purchase of the transceiver is prior to November 22, 1974, it may be used if it is either type accepted or the transmitted frequencies are crystal controlled. (*Type acceptance* indicates that the FCC has not necessarily tested a particular transceiver. The manufacturer has obtained FCC approval based only on test information supplied by the manufacturer.) After this date, all transceivers being marketed must be type accepted regardless of whether or not crystals are used. If your CB radio is not type accepted, note November 23, 1978, on your calendar. After this date, you are not permitted to use any transceiver that is not type accepted.

It is obviously very important to maintain the transmitted frequency within a specific tolerance. Excess drift can cause interference with an adjacent channel. On the 27-MHz band, frequencies must be kept within 0.005 percent of the specified frequency. For example, on channel 4 the center frequency is 27.005 MHz. If the transmitter is within tolerance, the broadcast frequency cannot vary by more than 0.0013502 MHz; that is, it must be between 27.0036498 and 27.0063502 MHz to be acceptable.

The percents of modulation and bandwidth are both carefully specified by the FCC, which recommends that frequencies up to 3000 hertz (Hz) be broadcast in order to maintain maximum intelligibility, with a maximum bandwidth of 4000 Hz on each side of the center frequency. No audio signal frequency should cause the transmitter section of the CB radio to be modulated more than 100 percent. The FCC has taken this one step further. Since May 24, 1974, every transceiver must have some type of limiting circuit so that all modulation is less than 100 percent. This latter specification is quite important, as overmodulation will produce a distorted signal and generate unwanted sidebands known as *splatter*.

The FCC has found that many of the limiting circuits now in use react too slowly. High-amplitude fast audio pulses may not be limited at all, and do overmodulate the radio-frequency (RF) signal. New regulations are under consideration to specify a minimum speed with which the limiting circuit must react to the audio signal.

Important Restrictions

Regulation of any type of broadcasting is important. Commercial broadcasting would be even more chaotic than it is without FCC rules. Limitations and rules of the game are even more important where CB radio is concerned, or there would be no fair play for most broadcasters. Here are some of the more outstanding regulations. Since you must have a copy of Part 95 concerning Citizens Radio Service from the FCC in order to get your license, you can read more details there.

1. Use your CB radio for communications only, not as a hobby. It should not be used for superfluous technical small talk, and you may not interfere with other transmissions.
2. Watch your language. Do not use obscenities or profanities.
3. CB radio may not be used for any illegal purpose.
4. Do not rent your facilities or use them to solicit sales of goods or services.
5. Communicate only with a specific station unless you are conveying information about an emergency.
6. Do not use CB radio to transmit music or other forms of entertainment. Do not use the signals for direct transmissions over a public-address system.
7. Do not transmit distress signals unless there is a true imminent danger.

Adding to Your Efficiency

Time is money. While you are on the road, you would naturally like to keep your conversation time down to a minimum and concentrate on your driving. First, identify yourself using your call letters. Next, use the recently revised code with 34 commonly used abbreviations shown in Table 1-2. Commit as much of this code to memory as you feel necessary; you will find that by using it your transmission time will be reduced considerably. Don't forget to repeat the call letters at the end of your conversation.

Table 1-2   10-Code Used by CBers for Brevity

10-1	Your signal is weak	10-18	Urgent
10-2	Your signal is good	10-19	In contact with ――
10-3	Stop transmitting	10-20	Location
10-4	Yes, OK	10-21	Call ―― by phone
10-5	Relay message to ――	10-22	Disregard or cancel
10-6	Busy	10-23	Arrived at scene
10-7	Out of service	10-24	Assignment has been completed
10-8	In service	10-25	Report to, meet
10-9	Repeat, say again	10-26	Estimated arrival time
10-10	Negative, no, denied	10-27	License and permit information
10-11	―― is on duty	10-28	Ownership information
10-12	Stand by, stop	10-29	Records check
10-13	Existing conditions	10-30	Danger, caution
10-14	Message, information	10-31	Pick up
10-15	Message delivered	10-32	―― units needed
10-16	Reply to message	10-33	Help me! Quick!
10-17	En route	10-34	Time

Note: A copy of the 10-code must be kept at the station and made available immediately to any FCC representative on request.
Numbers 10-35 through 10-39 have not yet been assigned definitions.

Codes can only be used if a copy and the meanings are kept on file at the station. It must be readily available to any FCC representative when requested.

As with most other communication outlets, interference may cause your signal to be poor. The receiver may have trouble understanding you. The phonetic alphabet in Table 1-3, which is recommended by the Road Emergency Associated Citizens Teams (REACT), could be quite useful in making yourself more intelligible.

As a matter of interest, General Motors sponsors REACT. It is a nationwide organization of CBers monitoring channel 9 to help and to advise people in need of aid. They frequently work with official organizations such as the police.

Table 1-3   International Phonetic Alphabet Recommended by REACT

A	Alpha	N	November
B	Bravo	O	Oscar
C	Charlie	P	Papa
D	Delta	Q	Quebec
E	Echo	R	Romeo
F	Foxtrot	S	Sierra
G	Golf	T	Tango
H	Hotel	U	Uniform
I	India	V	Victor
J	Juliette	W	Whiskey
K	Kilo	X	X-ray
L	Lima	Y	Yankee
M	Mike	Z	Zulu

Finally, the jargon. No matter what field you are working in, there is a specific language common only to it. If you don't know the language, you frequently find yourself standing by, and not comprehending the conversation. You may even be needlessly offended by something said because a specialized term may have a definition entirely unrelated to yours. CBers, too, have their own language. The following are definitions of the more commonly used terms.

Ace—an important CBer.

Advertising—a marked police car with the lights on.

Anchored modulator—the operator of a base station.

Apple—a CB enthusiast.

Back door—the last truck in a caravan.

Back down—reduce the speed of your vehicle.

Back out—stop transmitting.

Bad scene—an overcrowded channel.

Bear—any policeman.

Bear cave—a police station located on a highway.

Bear den—a police station.

Adding to Your Efficiency

Beast—a CB rig being serviced.
Be bop—a radio-controlled signal.
"Beer" tone—an intermittent tone.
Big daddy—the FCC.
Bleedover—hearing a signal from an adjacent channel.
Blessed event—a new CB rig.
Blood box—an ambulance.
Boast toastie—a CB expert.
Bootlegger—an unlicensed CBer.
Boy scouts—the state police.
Breaker—someone asking to use a channel.
BTO—big-time operator.
Bubble machine—police light.
Bubble trouble—tire is flat or other problem.
Bug out—to leave a channel.
Camera—police radar unit.
Cartel—a group hogging a specific channel.
Channel 25—the telephone.
Charlie—the FCC.
Clean—an area without any police units.
Cotton picker—a CBer who is not liked.
County mounty—a sheriff's deputy.
Cradle baby—a shy CBer afraid to ask someone to stand by.
Cub scouts—the sheriff's men.
Daddy-O—the FCC.
Dead pedal—a slow-moving vehicle.
Despair box—box with extra CB components.
Diesel digit—channel 15.
Don't tense—take it easy.
Drop the hammer—go as fast as you can.
Eatum-up—a roadside cafeteria or restaurant.
Eyeball—personal meeting with a CBer.
Feds—FCC inspectors.

Feed the bears—pay a speeding fine.

Fingers—a channel-hopping CBer.

Foot in the carburetor—police in pursuit.

Foot warmer—an illegal high-power linear amplifier.

Fox Charlie Charlie—the FCC.

Friendly Candy Company—the FCC.

Glory card—a Class D CB license.

Goon squad—a group hogging a channel.

Grab bag—illegal hamming by making a general CQ call to contact anyone around.

Green stamp road—toll road or road with radar, police, and/or weighing stations.

Green stamps—money for paying fines.

Hag feast—female CBers hogging a channel.

Hamster—a CBer who hams on a CB channel.

Handle—a nickname.

Happy number—an S-meter reading.

Henchmen—a group of CBers.

High gear—a linear amplifier being used.

Holler—call another station.

Hot pants—smoke or fire.

Hound men—police seeking CBers transmitting while in motion.

Land line—the telephone.

Linear—an illegal power amplifier to boost CB RF signal.

Local yokel—small-town cop.

Man in blue—policeman.

Move—go fast.

OM—a CBer.

Other half—the wife.

Over your shoulder—I am right behind you on the road.

Panic in the streets—area that is being monitored by the FCC.

Picture box—a radar setup.

Picture taker—a radar speed trap.

Adding to Your Efficiency

Plain wrapper—an unmarked police car; also indicates color of car.

Polit-tsei—the police.

Portable chicken coup—weighing station for portable trucks.

Prescription—FCC rules.

Rest-em-up—roadside rest area.

Rig—CB radio set.

Riot squad—neighbor with television interference.

Roger—OK.

Rollerskate—a passenger car.

Savages—CBers who are hogging a channel.

Shim—to illegally soup up a transmitter.

Shout—call a CB operator on the air.

Slaughter house—channel 11.

Slave drivers—CBers who hog a channel.

Slider—variable-frequency oscillator that permits illegal operation between channels.

Smokey—cop.

Smokey book—book showing police radio channels.

Smokey dozing—police car ready and waiting.

Smokey on rubber—moving police car.

Snooperscope—an illegally high antenna.

Sonnet—CBer who illegally advertises on the air.

Souped up—modified transceiver to deliver illegal high amount of power.

Splashover—signal leaking over to an adjacent channel.

Splatter—same as splashover.

Stroller—CBer using a walkie-talkie.

Struggle—trying to get to talk on a channel.

Sucker—a CB rig on the service bench.

Sunbeam—a witty CBer on the air.

Thin man—weak carrier.

Thread—wires used on CB rigs.

Ticks—minutes going by; each tick is 1 minute.

Tijuana taxi—a marked police car.

Tin can—a CB transceiver.

Tooled up—souped-up rig.

Train station—traffic court known to fine all violators.

Twin pots—CBer with two transceivers manufactured by the same firm.

Wall to wall—very strong CB signal.

Wrapper—color of a car, including police car or truck.

WT—walkie-talkie.

XYL—CBer's wife (ex-young lady).

YL—Young lady, a single Ms.

2

Transceiver

Taken as a whole, any piece of electronic equipment appears to be extremely complex. Fortunately, the CB transceiver (more commonly known as the radio) can be subdivided into a number of simple sections. None of these sections is particularly strange to the service technician.

Basic Radio

No electronic gear works without a source of power, whether it be a solar source, a storage battery, or power available at a wall outlet. Mobile CB radios commonly obtain power from the storage battery in the car or truck. Base station units may have built-in circuitry to convert ac power available at the wall outlet to dc for circuits in the transceiver. This dc must be well filtered to keep hum out of the transmitted signal and to reproduce clean audio at the loudspeaker output from the receiver section of the transceiver.

The section of the transceiver designed to receive the signal is not unlike the portable radio commonly used for receiving standard broadcasts. The receiver circuit in both the transceiver and the portable radio consists of an RF amplifier, a local oscillator, an intermediate-frequency (IF) section, and a detector. All sections work together to convert the received signal to electronic audio energy peaks. Finally, there is an audio amplifier to increase the strength of these electronic audio signals so that they can be fed to a loudspeaker to be transformed into audible speech.

In the transmitter section of the CB radio, an RF oscillator and RF amplifier generate and amplify the carrier signal. It is coupled to an antenna and radiated into the atmosphere so that it can be received by the idling receiver. In amplitude-modulated (AM) transmitters, an RF signal by itself does not furnish intelligence. The RF must be amplitude modulated by the audio signal so that the RF carries information to the listener. The audio amplifier thus serves a dual function in the CB radio. In the receive mode, it amplifies audio pulses so that audible sound is reproduced in the loudspeaker. When connected as part of the transmitter, the audio amplifier magnifies the weak audio signal pulses from the microphone and applies it to the RF amplifier or power output stage of the transmitter for modulation purposes.

Finally, there is the switching circuit, which changes the operating mode of the transceiver from receive to transmit. In the process, the audio amplifier is switched from its function of reproducing received signals (after detection) to amplitude modulating the RF. On some transceivers, the switch is also used to move the antenna from the receiver input circuit to the transmitter output. Figure 2-1 depicts this.

Three sections of a double-throw slide switch are shown in the receive (REC) position. If you trace the circuit, you will find the antenna connected to the receiver RF and detector stages through section S1A of the slide switch. The detected audio output is connected to the audio amplifier through section S1B of the switch; section S1C is used to connect the amplifier output to a loudspeaker. If the switch were set in the alternate position, you would find the microphone preamplifier connected to the audio amplifier input through section S1B of the switch, while section S1C is used to connect the output from the amplifier to the transmitter for modulation purposes. The antenna is connected to the output of the transmitter circuit through section S1A of the three-pole switch.

All the switching shown in Fig. 2-1 is in the transceiver. Switching is normally not accomplished by a device mounted on the panel of the CB radio, but by a switch located in the microphone. When transmitting, the push-to-talk lever on the microphone is pressed. In Fig. 2-2, it is shown as capable of closing the contacts of the switch in the microphone. This

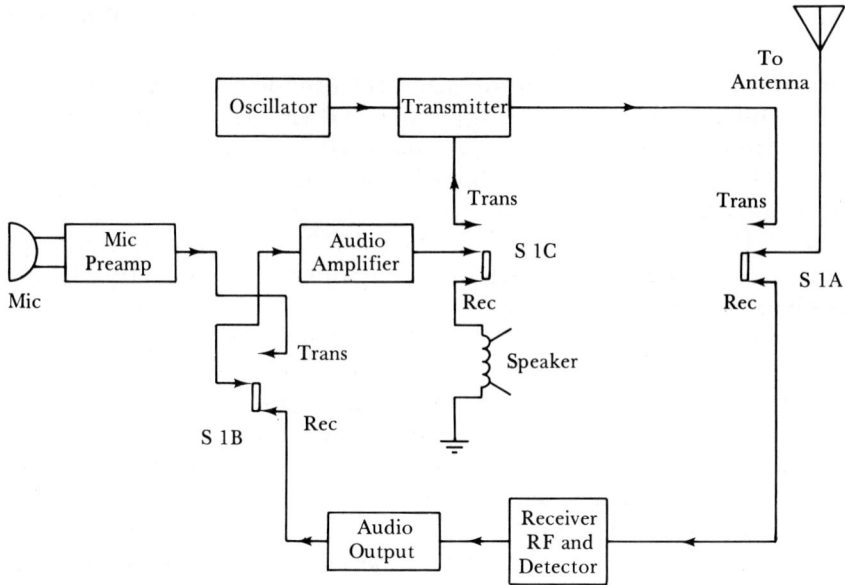

FIG. 2-1 Basic CB transceiver circuit.

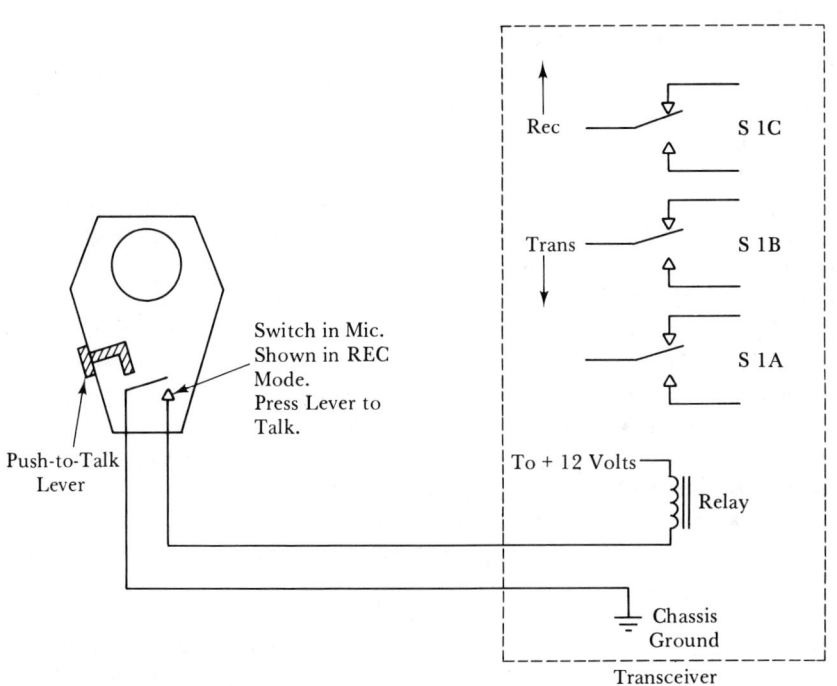

FIG. 2-2 Relay in transceiver can be energized by push-to-talk lever in MIC to change from receive to transmit mode.

switch completes a circuit to energize a relay in the transceiver, which, in turn, shifts the settings of the three-pole TRANS-REC switch in the CB radio from the receive to the transmit mode.

The switch in the microphone is normally open. In this condition it does not affect the receive and transmit circuitry in the transceiver. One side of the relay is connected to a +12-volt (V) supply, but the other side is connected to the open switch in the microphone. As the relay circuit is not complete, the center arms of the switches (mounted on and activated by the relay) are in the REC mode, completing RF and audio circuits similar to those shown in Fig. 2-1. Upon pressing the push-to-talk lever in the microphone, the current path for the relay is completed through the switch by connecting the unused lead from the relay to ground on the transceiver chassis. Twelve volts is across the relay. The magnetic field due to the voltage across the relay coil (or more precisely the current through the coil) forces the center or moving wiper arms of the three switch sections on the relay toward the magnet. Connections of the wipers to the various upper contacts are broken. The wiper arms now make contact with their respective lower terminals or contacts. The REC circuit is deactivated while the TRANS circuit in the CB radio is now complete through these wiper arms.

Relay switching was and still is commonly used. Quieter electronic switching is usurping much of the popularity still enjoyed by relay switching. A basic circuit of transceivers using electronic switching is shown in Fig. 2-3.

All switching functions are performed by the double-pole double-throw (DPDT) remote slide switch in the microphone. Although shown as a slide switch for convenience, and although this type of switch predominates in this application, leaf-type switches have also been used. Despite actually being in the microphone, the switch is shown here for clarity as being in the CB radio proper. Arrows from the microphone to the switch indicate the correlation between the true physical location of the switch in the microphone and its functional location in the circuit.

In this particular arrangement, a DPDT switch is shown. The antenna is permanently connected to both the input of the receiver and the output of the transmitter through some type of passive isolating or matching circuit. With the switch as shown, section 1 has lifted the transmitter circuit off ground. Although B+ from the power supply is still connected to the circuit, the negative side of the supply has been removed. There is no voltage across the transistors in the transmitter section. It cannot function. The microphone preamplifier electronics is similarly floating off ground, and is thus also out of the circuit in the REC mode. The common lead connecting the microphone preamplifier to the transmitter should not confuse anyone, as the connections refer

Basic Radio 17

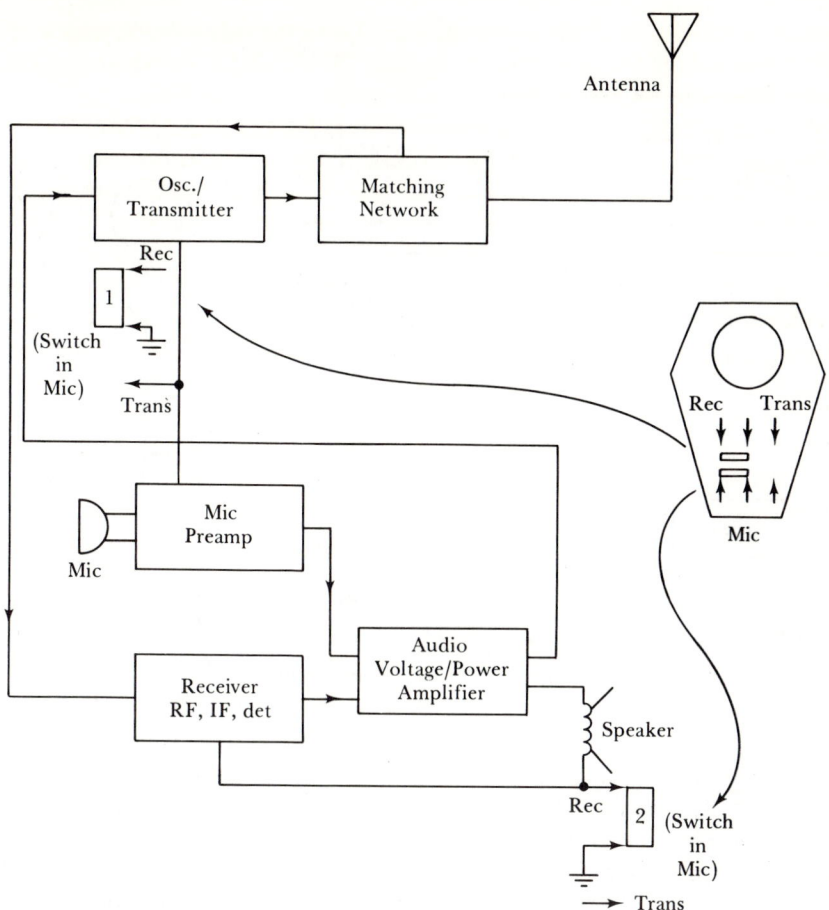

FIG. 2-3 Circuit employing electronic switching.

only to the power supply leads and not to the functional **audio** and **RF** circuits.

Tracing the circuit in the receive mode, the antenna is **permanently** connected at the input to the receiver. The minute audio output from the receiver section is amplified and applied to the loudspeaker. The receiver electronics and the loudspeaker are returned to ground through section 2 of the switch, so that signal can be received and the audio can be reproduced in the REC mode of operation. At the same time, the transmitter section is disabled by section 1 of the switch.

As for the transmitting function, the loudspeaker and receiver RF sections must be disabled when in the TRANS mode of operation. This is accomplished by section 2 of the switch. When pressing the push-to-

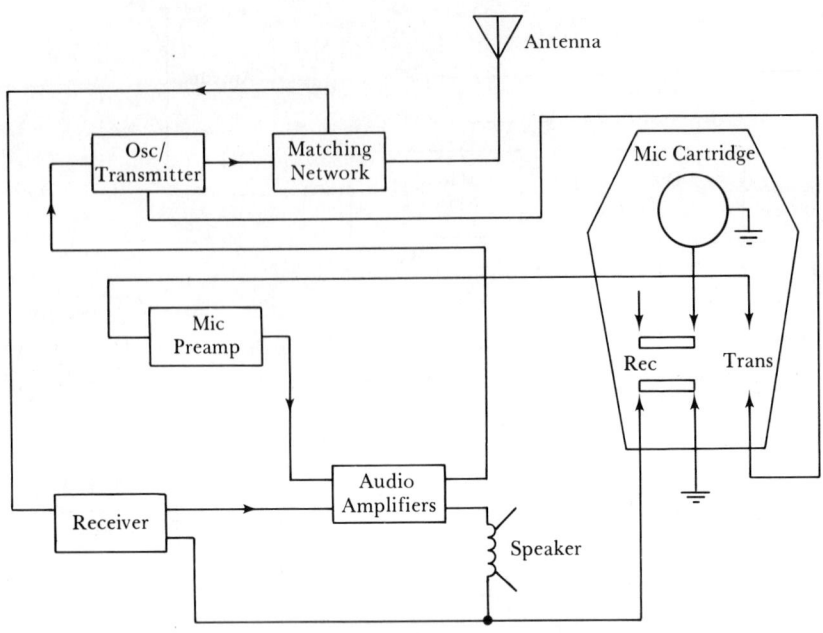

FIG. 2-4 Conventional arrangement for electronic switching.

talk button on the microphone, both the speaker and receiver electronics are lifted off ground. Meanwhile, section 1 of the switch is performing its function by connecting the transmitter and microphone preamplifier sections to their negative ground returns.

Audio signals from the microphone are amplified by the preamplifier and audio amplifier sections. Magnified signals are applied to the now functioning transmitter section for modulation purposes. Being connected to the antenna, the signal from the transmitter is radiated into the atmosphere. No signal from the audio amplifier is reproduced through the loudspeaker as one of its leads has been lifted off ground at section 2 of the slide switch.

Many transceivers utilize the circuit as described, but variations can be found everywhere. The two functions of the switch are frequently performed by one single-pole double-throw section of the DPDT switch, as shown in Fig. 2-4. Here the center lug is grounded while the speaker and receiver are connected to one of the remaining lugs on this switch section, and the transmitter is connected to the other lug. At the same time, the second section or pole of the switch is used to break the connection from the microphone cartridge to the preamplifier so that it does not have to be lifted off ground.

Transceiver Features

Basic transceivers have been described. Very few, if any, CB radios are as void of refinements and enhancements as the units discussed above. Some of these appendages are extremely useful, and no self-respecting unit would be without one or more of these.

Automatic Volume Control

Just about every broadcast band and all-wave radio made has automatic volume control (AVC), which is also included on every self-respecting CB radio. Because of this feature, you can switch or tune to different stations (channels) without being forced to readjust the volume control each time.

AVC can readily be explained with the help of Fig. 2-5. The receiver shown is similar to most superheterodyne radios. The audio-modulated RF is received and then magnified by the RF amplifier. This stage is omitted on many receivers. Instead AM radio signals are fed directly to the mixer. Here, signal from a local oscillator is mixed with, or beats against, the received RF. The frequency of the local oscillator signal is usually 455 kilohertz (kHz) higher than that of the incoming signal.

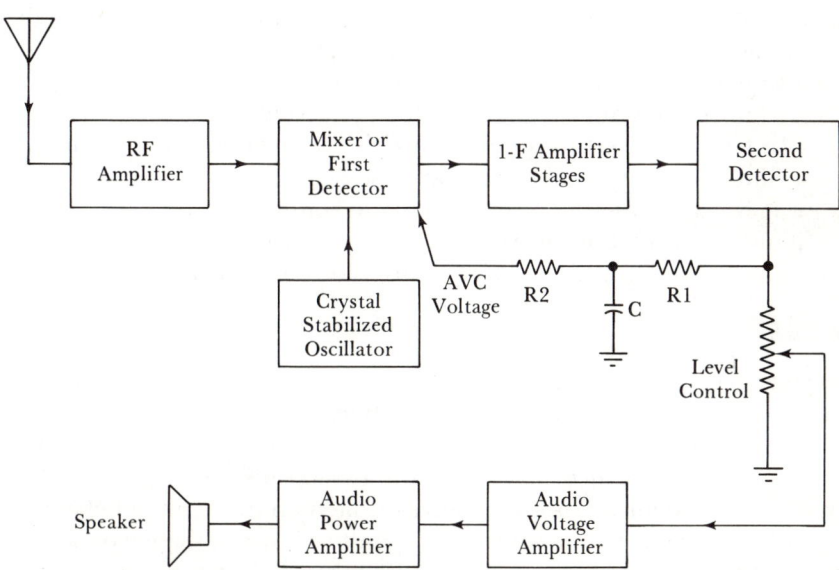

FIG. 2-5 The receiver-single conversion.

On CB radios, tuning of the receiver is not continuous. Following the practice common to television, each station is usually selected by setting a switch to a specific position. Because of this, the local oscillator frequency is critical. Oscillator frequencies in CB radios are normally set and stabilized by crystals in the circuit.

Due to the mixing function, four frequencies are at the output of this stage. These are the received and modulated RF, the oscillator frequency, the modulated sum of the oscillator and RF frequencies, and the modulated difference between the oscillator and RF frequencies. It is this latter modulated difference frequency, usually 455 kHz, that is selected and amplified by the IF stages.

After having passed through the IF amplifiers, the signal is detected. The output from the detector is audio. It is applied across the level control, from where it is fed through an audio voltage amplifier and power amplifier to the loudspeaker.

Now let us return to the audio at the top of the level control. Besides being fed to the voltage amplifier for audio signal reproduction, the audio is filtered by the R1–C network and fed back through R2 as a variable bias voltage for the earlier RF and/or IF stages. As the feedback voltage varies, it changes the bias on transistors in the RF and/or IF stages. The gain of these devices is, in turn, varied by the amount of voltage applied to the bias circuits. The filtered audio voltage, fed back for dc bias control, is the AVC voltage in question.

When filtered, the AVC voltage is dc, the amplitude of which depends upon the strength of the received signal. If the amplitude is high, the voltage fed back is such as to reduce the gain of the RF and/or IF circuits; if low, the gain of the circuits is increased. Thus the output level is maintained constant by this AVC feedback voltage.

In television, picture signals must also be maintained at a constant level. Similar gain-control circuits are used there for this purpose. Because it is not audio, the term "automatic volume control" would be a misnomer. Instead, the more universal name *automatic gain control* (AGC) is used. AVC in CB radios is frequently referred to as AGC.

Squelch

Receiver sections of CB radios have tremendous gain. Because of this, there is an enormous amount of noise when no signal is being received. As excess noise from an idling receiver is objectionable, a squelch circuit is practically always incorporated in the basic transceiver. The squelch circuit can be regulated using a control located on the front panel of the CB radio. You can set the control to a point at which no signal or sound will be reproduced through the loudspeaker unless the received RF signal is above a predetermined threshold level.

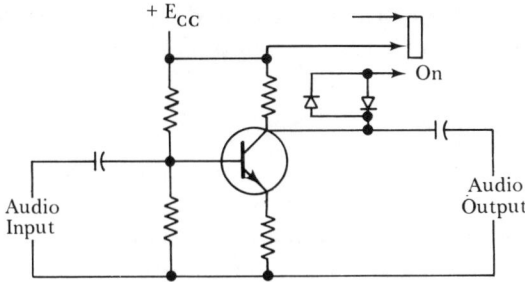

FIG. 2-6 Noise limiter.

The squelch circuit function on most transceivers is based on information supplied by the AVC voltage. It is fed as direct current to the base or emitter circuit of the first audio amplifier stage in the receiver, keeping that stage cut off until the received signal is above a specific level. Once the minimum required signal level is exceeded, the affected stage is turned on, allowing the audio signal to pass through to the loudspeaker.

Another type of squelch circuit receives and amplifies the noise. It is detected and fed to the audio voltage amplifier stage as a dc bias voltage. Bias voltage is in such a direction as to turn the stage on or off, depending upon the amount of received or circuit noise available. We need not concern ourselves too much with this system, as it is seldom used in CB radios.

Automatic Noise Limiter

The automatic noise limiter (ANL) circuit is usually activated by a switch on the front panel. Functionally, the ANL blocks the audio at the second detector from passing to the level control (see Fig. 2-5) when signal is accompanied by large noise pulses. These pulses do not necessarily have to be caused by noises in the air surrounding the earth. Pulses may be due to noise from an automobile's ignition system or to electric devices used in the home, including some fluorescent fixtures. Audio signal blockage is only for the instant that noise pulses actually exist.

Besides systems that do not allow signals with excess noise to pass through the audio circuit, there is the clipping device. For example, assume that the peak audio signal ever present across a load in the collector of a transistor is 0.6 V. If you should place two silicon diodes in parallel across this load with the cathode of each one connected to the anode of the other, as in Fig. 2-6, they would limit the voltage across the load to the required maximum 0.6 V in both directions. Positive and negative peaks in the signal due to noise would be limited. This is due to the fact that a forward-biased silicon diode will break down and appear as a shorted or low-value resistor when more than 0.6 V is across it.

Radio-Frequency Gain Control

In strong signal areas, the received signal may overload the input stage. Once this happens, the audio output is likely to be distorted. One way of overcoming this problem is to keep a substantial distance between the transceiver receiving the signal and the transmitter producing the strong RF. This is seldom a possible solution. To overcome this difficulty in a practical manner, the gain of the RF stage must be reduced so that it will not be overloaded by a strong signal. A special variable gain control has been added to the front panel to perform this function. Although not commonly found on CB radios, the value of an RF gain control in many instances is undeniable.

Dual Conversion

Many advertisers of CB sets boast that the transceiver has a dual conversion receiving system rather than the single conversion system in Fig. 2-5. The primary advantage of a dual conversion system is to improve selectivity by limiting the bandwidth of the receiver and in this way minimize interference from signals not on the band. A block diagram of this system is shown in Fig. 2-7. It is similar to the single conversion receiver described under Automatic Volume Control.

    Audio-modulated RF is amplified by the RF stage after the existence of the signal has been sensed by the antenna. The first, or high-frequency (HF), oscillator beats with the RF in the HF mixer stage. Of the four frequencies at its output—RF, oscillator, sum, and difference—the modulated difference frequency (usually about 10.6 MHz) is selected by the high IF circuit. It is then fed to the low-frequency (LF) mixer, where this modulated 10.6 MHz signal beats against a signal generated by the LF local oscillator. The modulated difference frequency, usually 455 kHz, is selected from the four at the output of this mixer, and passed through the low IF amplifier. After being detected, the resultant audio is amplified and made audible through the transducer—the loudspeaker.

Delta Tune

Although crystal controlled, some transmitters will produce signals that are slightly off the assigned frequency. Some receivers have limited front panel tuning facilities to allow you to shift the receiver oscillator frequency to some degree. This provides the flexibility of tuning accurately

Transceiver Features

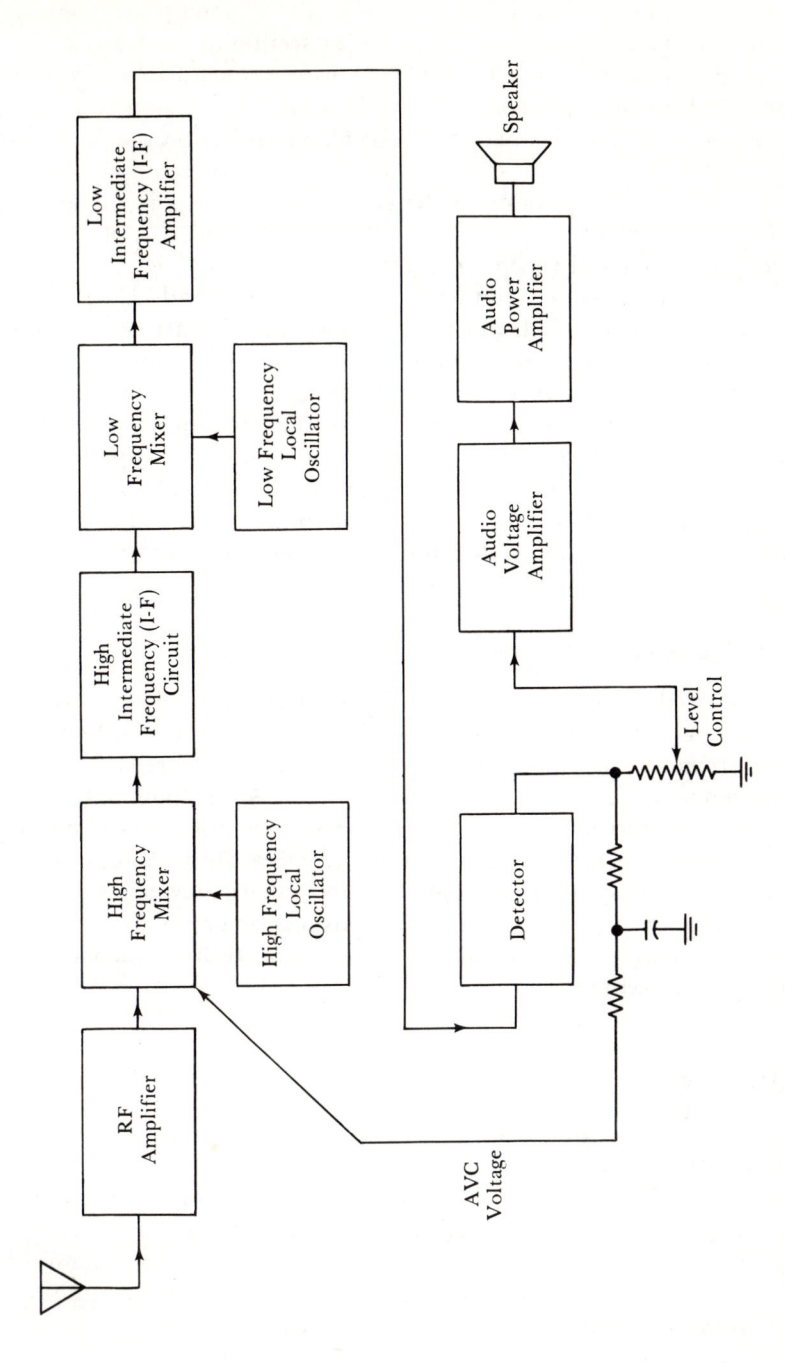

FIG. 2-7 Dual-conversion receiver.

to the transmitted frequency, even if it differs somewhat from the assigned value. The facility on the receiver section of the transceiver may be in the form of continuous tuning where you can change the frequency of the second local oscillator in a dual conversion receiver. The frequency can be varied about 1.5 kHz above and below the center value.

More often, a three-position switch is used to accomplish this function. At the center position of the switch, components in the local oscillator circuit are arranged so that the signal supplied is at the proper frequency for good reception when the transmitted signal is exactly at the assigned frequency. Inductors or capacitors are added to the circuit in the alternative switch positions. These components change the frequency of oscillation. When one component is in the oscillator circuit, the signal that is 1.5 kHz *below* the center frequency can be received. In the alternative setting of the switch, a different frequency-determining component is placed in the circuit. Now oscillation is proper for reception of a signal 1.5 kHz *above* the center frequency. The switch is set to the position that provides the best signal-to-noise ratio at the output of the receiver and/or the maximum reading on the S-meter.

Automatic Channel 9 Monitor

As discussed in Chapter 1, channel 9 is for emergency use only. In the vicinity of an emergency, where you would either like to help or just avoid trouble, it would be helpful if you were forewarned of the emergency. Obviously, you are not always tuned to channel 9 to get this information and may be using your transceiver for other communications functions. Facilities on some transceivers alert the operator that channel 9 is being used and that a possible emergency exists. For example, a lamp may glow when there is transmission on channel 9, or other indicators may be used to make the operator aware of activity on the emergency channel.

S-Meter

An S-meter is provided on the front panel of most transceivers to indicate the relative strength of the received signals. Although not an absolute necessity, it is a useful indicator that the received signal on any channel is of sufficient strength to be intelligible. Larger S-meter readings indicate stronger signals are being received. The S-meter also works hand in hand with the delta tune feature, as the channel is tuned properly when the pointer deflection on the meter is at a maximum.

S/RF Meter

In the receive function, the S/RF-meter behaves as does the simple S-meter just described. Its primary advantage over the S-meter is when transmitting. Relative RF power from the transmitter section of the transceiver is indicated on the meter. Because it is a fixed value, full power output is usually indicated by a full-scale or near-full-scale deflection of the pointer. A radically lesser deflection should alert you to a defect in the CB set. The most vulnerable component in the transmitter circuit is the RF output transistor feeding the antenna.

Modulation Lights

Because microphones are electromechanical devices, they are most vulnerable to breakdown. A wire may break in the cable. The junction between the cable and connector to the transceiver is another weak point. Finally, any switch or circuit in the microphone proper may go bad. Because of these factors, you may find yourself talking into the microphone, but there is no output. RF output signals at your transceiver will therefore not be modulated.

Many CB radios have *modulation lights* to indicate if audio from the microphone is reaching the transmitter section. These may be simple independent lights visible through a jewel in the panel, or they may be mounted in and observed through the S-meter. They flash when you speak through the microphone if all is well with the modulation.

Warning Lights

Transmitter output stages must be loaded properly by an antenna when transmitting. If the antenna is shorted or open, not only will it prevent you from "getting out," but it may cause the power output transistor in the transmitter RF circuit to fail. Some transceivers have lights to warn you that the antenna has not been connected properly.

SWR Meter

The standing-wave ratio (SWR) is an indication of how well the antenna is matched to the impedance of the coaxial cable that connects it to the transceiver. Antenna impedance must equal the impedance of the coaxial cable, usually 52 ohms (Ω), if the maximum power is to be transferred from the CB rig to the antenna system. SWR meters used to perform this measuring function are normally accessories. In a few transceivers, an

SWR measuring instrument is included as an integral part of the RF output circuit.

Single Sideband

Single sideband (SSB) is not a "feature" in the sense of the word as it has been applied up to now. Instead, SSB is a variation of the 23-channel AM carrier CB broadcasting system. SSB transmission provides the flexibility of effectively doubling the number of channels available to the licensee.

The conventional AM signal used on the CB band can be shown to consist essentially of three items: (1) the carrier or RF on which the audio rides, (2) an upper sideband, and (3) a lower sideband. The sidebands occur only when audio is modulating the RF carrier. The frequency of the upper sideband is the RF plus the modulating audio frequency; that of the lower sideband is the RF minus the modulating audio frequency. The amplitude or signal strength of the sidebands depends upon the magnitude of the audio signal modulating the RF.

Individual sidebands may be transmitted while the RF carrier is suppressed. Circuits used in some CB transceivers allow one sideband at a time to be transmitted. Because there is an upper and lower sideband for every channel, two independent signals can be received simultaneously on each channel, one through each sideband. They are separated by filters or other circuitry. The number of available CB outlets can thus be increased from 40 to 80.

Added Features

Up to this point, features that are useful in the actual process of receiving and transmitting have been described. At very little extra cost to the manufacturer, a number of features relatively unrelated to the primary function of the transceiver have been added.

For example, there is an audio amplifier in the CB radio. With just a little switching, it can be converted for use as a public-address (PA) microphone amplifier. To embellish this and the tone quality (using the words "tone quality" very loosely) of the reception, a simple treble cut tone control may be supplied on the front panel. In order not to be limited to the small loudspeaker usually mounted in the transceiver, jacks are supplied on the rear panel to provide connecting facilities for a remote loudspeaker or headset. When you plug into one of these jacks, you disable the local loudspeaker in the transceiver, and the audio output from your CB radio is connected to the remote unit in both the CB and PA modes of operation.

Microphone

Although separated from the transceiver proper, the microphone and antenna must be considered as integral parts of the CB rig. They deserve special attention here, because if these two items are weak links the sophistication of the CB radio itself is unimportant. The microphone will be described here and the antenna in the next section.

There are two basic types of microphones, the dynamic and the ceramic. Dynamic microphones usually have wider frequency reproducing capabilities than do ceramic types. However, with modern technology, one can be made as good as the other. Because the audio frequency range of the overall rig is limited from about 300 to 3000 Hz, performance quality can be just as satisfactory regardless of the type of microphone used with the rig. It should be noted, however, that each CB radio is designed for use with one, and only one, of the microphone types available.

A dynamic microphone consists basically of a coil that can move in a magnetic field. The coil is physically connected to a large diaphragm. As you speak into the microphone, you alternately apply pressure to and remove pressure from the diaphragm. It moves, and this movement is synchronized with your speech. Because the diaphragm is attached to the coil, the coil also moves in time with your speech. Current is generated in a wire or coil of wire when it moves in a magnetic field. Voltage appears across the coil owing to the current flowing in the wire. These voltage peaks are an exact electrical replica of the speech that initially caused the diaphragm and coil to vibrate. Coupled to a voltage amplifier in the CB radio, the generated voltage peaks are magnified sufficiently by the electronic circuit to be capable of modulating the RF in the transmitter section.

Because a coil is an inductor, it has an impedance, which normally ranges from about 50 to 1000 Ω. More voltage is usually generated by coils with higher impedances than by those with the lower impedances. Manufacturers generally use microphones with impedances specified at about 500 or 600 Ω.

Ceramic or crystal microphones usually provide more voltage at their output terminals than do their dynamic counterparts. A ceramic type of cartridge consists of a solid slab of a special material that generates a voltage when it is forced out of shape. As in the dynamic microphone, a diaphragm is used. Here it is mechanically connected to the slab. As speech hits the diaphragm, it, in turn, causes a force to be applied to the slab.

Characteristics of ceramic and crystal cartridges are similar to those of capacitors rather than inductors. Impedances are high. Because of this they are not rated by their impedance, but rather by the capacity of the slabs. Capacity normally ranges from 500 to 1500 picofarads (pF), with 750-pF cartridges being most popular with manufacturers. Cartridges have been made with capacitances of up to 7000 pF, but these are more the exception than the rule.

High-impedance cartridges require the amplifiers that they feed to have input impedances ranging from 100,000 Ω to 3 megohms (MΩ). Hence microphones with ceramic cartridges cannot be used with the low input impedance amplifiers normally designed for use with low-impedance dynamic cartridges. Similarly, dynamic cartridge microphones cannot normally be substituted for ceramic or crystal types. Output voltages generated by the latter types are higher than voltages supplied by dynamic cartridges. Therefore, amplifiers built into transceivers that are to be used with dynamic microphones have higher gain than those designed for use with ceramic types. Sufficient gain is usually not available to use a dynamic microphone as a substitute for a microphone with a crystal or ceramic transducer.

Ceramic and crystal cartridges have been referred to here as if they were identical devices or behave in the same manner. Basically, they do behave identically. Ceramic types are favored because crystal cartridges are more susceptible to destruction by high humidity and temperature. From here on, we shall refer only to ceramic cartridges, but it should be remembered that many of the characteristics are also true of crystal types.

Microphone manufacturers usually supply a specification indicating the output voltage from the microphone at 1 kHz. Ratings are either in millivolts (mV) out or in a negative number of decibels (dB) below a specific level. When rated in millivolts, it is obvious that the microphone with the higher voltage rating will provide more output signal. When in decibels, smaller numbers (disregarding that the numbers are negative) indicate the device with the higher output. To compare ratings, all specifications should use the same acoustic audio input at 1 kHz as a standard—normally 10 microbars (μbar). Microbar is a quantity of audio or acoustic pressure applied to the microphone element.

Besides basic microphones housing only the ceramic or dynamic cartridge, CB transceiver and accessory manufacturers have also made available microphones with built-in amplifiers. These are known as *power microphones*. To better understand their function, it is necessary to observe the transmitted RF signals with and without modulation, as shown in Fig. 2-8.

The series of high-frequency sine waves shown in Fig. 2-8a represent the unmodulated or pure RF at about 27 MHz generated in the transmitter section of the CB radio. Figure 2-8b shows a cycle of low-

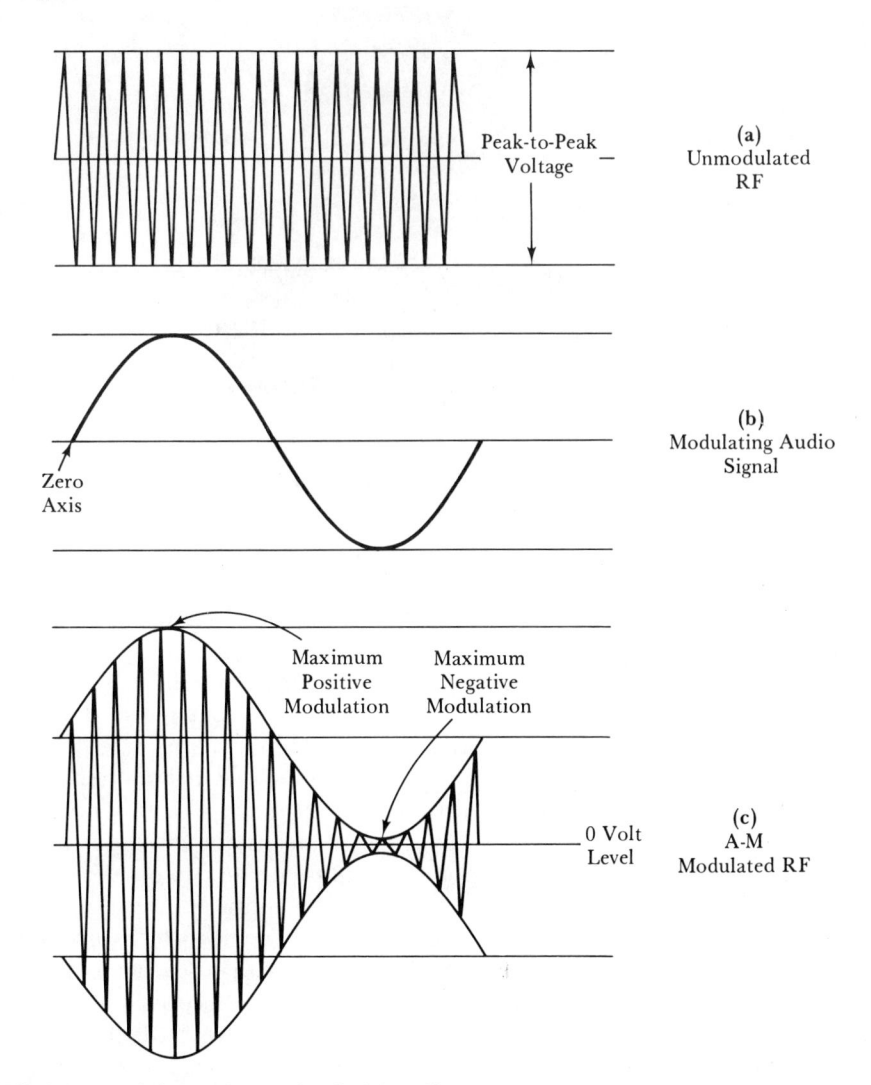

FIG. 2-8   Modulating the RF signal with audio.

frequency audio that is used to modulate the RF. The composite signal, with the audio modulating the RF, is shown in Fig. 2-8c.

The quantity of power delivered to the antenna system depends not only upon the amplitude or size of the available RF signal, but is also related to the magnitude of audio riding on the RF. Power output increases with the amplitude of the modulating signal. Note the point in Fig. 2-8c marked **max negative modulation**. At this point, the size of the audio signal is at its maximum, while the magnitude of the RF is just

FIG. 2-9 Power microphone with readily accessible level control. (Courtesy Mura Corp.)

about zero. This is one point in the audio cycle at which the RF is modulated 100 percent. A second point at which modulation is 100 percent is marked **max positive modulation**. Here the RF is double its unmodulated amplitude.

Microphones must supply enough voltage so that there is sufficient audio to modulate the RF very close to 100 percent if the delivered power at the output is to be at a maximum. To assure this, microphones with built-in amplifiers are supplied, should a CBer desire one.

Some power microphones (see Fig. 2-9) have a readily accessible audio gain control. It provides a convenient means of easily adjusting the audio output from the microphone and, in turn, the percent of modulation. Other power microphones have no gain controls. Amplification is either fixed or the amount of gain may be selected in steps by simply removing specific resistors (provided for this purpose) from the amplifier printed circuit board inside the microphone proper.

FCC rules dictate that a limiter be built into every CB transceiver. Modulation is thereby limited to just under 100 percent, regardless of the magnitude of the audio signal applied. Because of this, audio output, even from high-gain power microphones, will not be able to modulate the RF more than 100 percent. But power microphones can supply enough signal so that modulation is almost 100 percent for more than just the two individual instances in the sinusoidal cycle. More power can then be delivered to the antenna for a longer period of time than is possible with microphones encompassing only a bare dynamic or ceramic element. It is true that some distortion is generated in the process owing to clipping or limiting, but because audio bandwidth is narrow, the distortion is barely noticeable.

Antennas

Although it appears to be a rather simple component, the antenna's behavior and operation are probably the most complex and least understood of all parts in the rig. Most aspects are well beyond the scope of this book. It should be noted, however, that the antenna is frequently the weak link in the entire system.

Half-wave antennas are most desirable. Considering that the CB band is around 27 MHz, the antenna should be about 18 feet (ft) long. This number is approximate in every sense of the word, as the length also depends upon antenna height above ground and the fact that the 18-ft dimension is a calculation of a half-wavelength in air. A half-wavelength in the metal comprising the antenna is about 5 percent less than 18 ft. Whatever the actual required length, the half-wave antenna as calculated is too long for use in any practical mobile installation.

The length can be reduced by half when vertical antenna rods are used. Mounted vertically, only a quarter-wavelength is needed above ground. The other portion is a quarter-wavelength reflected by the actual ground or by a large metal surface, such as the roof of an automobile. (When talking about antennas, the large metal surface is referred to as a *ground plane*.) The full half-wave antenna requirement is thus satisfied.

But even a quarter-wavelength rod is too long to be practical in most installations. A shorter rod is frequently used. A coil, or inductance, is placed in series with the rod. An inductor in series with an antenna rod makes it appear as if it is electrically longer than its physical size. (Obviously, a capacitor in series with an antenna rod makes it appear to be electrically shorter than its physical size would indicate.) These inductors are known as *loading coils*. When an antenna mast is shorter than a quarter-wavelength, a loading coil is placed at the base or center of the antenna to make it appear electrically longer than it actually is.

Antennas also have impedances. For a maximum transfer of power from the transmitter to the antenna, the impedances must be equal to that of the shielded lead connecting the antenna to the transceiver. An SWR reading indicates just how good this match really is. The length of rod or mast can be adjusted for a maximum transfer of power (and minimum SWR) when a particular loading coil is used.

Different types of antennas are readily available. Some are to be mounted on the rain gutter of the car. Others are mounted on the automobile trunk or roof. Base station antennas are available for roof or tower mounting.

Two antennas located on opposite sides of the car or truck cab are sometimes used simultaneously and fed as one to the transceiver. Signals from both must be in phase. The primary advantages of this *diversity reception*, as it is known, are to eliminate the effects of fading and some of the directional characteristics of the transmission and reception. At the same time, there is an effective increase in the broadcasting range and the reception sensitivity of the rig.

Some systems use several independent antennas mounted at different areas on the car or truck. A switch is used to select the one providing the best performance in a particular situation. A discussion of antenna types, performance variations, installation, and servicing details is reserved for Chapter 9.

3

Accessories

Dictionary definitions of the word "accessory" are "1. a subordinate part; 2. something added for convenience." Considering the overall CB rig, the microphone and antenna may be thought of as accessories defined by "subordinate parts." Are they really *subordinate*? CB transmission and reception cannot be realized without a microphone and antenna. They are factually essential. There are many design variations on these items that "gild the lily," but in all installations one microphone and one antenna are *must* items. Because of their importance, separate chapters are devoted to the microphone and antenna.

This chapter deals with true accessory items, those described by the second definition of the word. These items may be "added for convenience" to the rig, but some of them also help improve or optimize performance. Only legitimate types of accessories are discussed here. Many gimmicks and gadgets designed to attract your attention are just impulse shopping items and serve little or no function. Although these items are not considered here, the reverse is not true, and we apologize to designers and manufacturers for oversight or omission of useful items.

Standing-Wave-Ratio/Power/ Field-Strength Meters

The SWR/power/field-strength meter is used when adjusting the antenna system and when measuring the transmission characteristics of the transceiver. It would seem logical to delegate the discussion of the meter to a chapter dealing with instruments. But it is more functional than being merely a "tester." Once various adjustments have been made using the instrument, it may be removed from the antenna transmission line and stored in a tool box to rust until a new antenna is to be installed and checked. It is, however, more likely to be left connected in the antenna circuit to continuously monitor the SWR and output power, assuring the operator of optimum performance at all times. Of all accessories to be described, this is probably the most essential one.

RF signal generated in the transmitter is fed to the antenna through a cable known as a *transmission line*. If the impedances of the transmission line and antenna are identical, all the signal that can be transferred to the antenna is at the antenna. It can then perform its prime function of radiating RF into the atmosphere. Should impedances differ, a portion of the signal is reflected back to the transmitter from the antenna. The quantity reflected back depends upon how unlike the impedances are. The more they differ, the larger the reflected signal.

Forward-flowing signals from the transmitter to the antenna and reflected signals from the antenna to the transmitter add on the transmission line. The sum appears as a stationary signal on the line and is known as a *standing wave*. Standing waves are radiated from the transmission line into the air, but very poorly. Transmission lines are not designed to radiate signals. Only the antenna should be used for this purpose, as it is the most effective and efficient radiator. Standing waves on transmission lines must be minimized.

Standing waves have voltage and current minimums and maximums. The ratio of these maximums to minimums is known as the *standing-wave ratio* (SWR). It may also be considered as the ratio of the forward (FWD) to the reflected (REF) voltage, or the ratio of the two impedances (transmission-line impedance and antenna impedance) involved. However you wish to look at it, all ratios produce the same SWR reading. Antenna length should be adjusted for a minimum SWR.

SWR is an indication of how much power is lost in the process of transferring it from the transceiver to the antenna. The percent of

Table 3-1   Percentage of Power Lost When SWR Differs from a 1:1 Ratio

SWR	% Power Lost	SWR	% Power Lost
1.0:1	0.		
1.1:1	0.227	2.5:1	18.4
1.2:1	0.827	3.0:1	25
1.3:1	1.71	3.5:1	30.9
1.4:1	2.78	4.0:1	36
1.5:1	4	4.5:1	40.5
2.0:1	11.1	5.0:1	44.5

power lost for various SWR readings is shown in Table 3-1. One-fourth of the power from the transceiver that would ordinarily reach the antenna never gets there if the SWR is 3:1. Assuming that the transceiver being measured is capable of delivering the legal 4 W to the antenna, only 3 W will reach it if the SWR is 3:1. This is quite a large loss of power.

Output power readings on these instruments are accurate only if the SWR of the system is 1:1. All necessary adjustments must be made to minimize the SWR before measuring power or the readings will be maccurrate. Accuracy increases as the SWR decreases.

To measure SWR and power, connect the antenna jack on the transceiver to the transmitter input terminal on the instrument using a short length of matching coaxial cable. Connect the cable from the antenna to the antenna output jack on the SWR/power meter. The instrument should not be removed from the circuit after measurements have been made, but should be left connected at all times to monitor power and SWR. This can be done successfully as the instrument's presence in the circuit does not materially affect the matching or absorb power. It must, however, be removed from the circuit when making field-strength measurements.

A typical instrument is shown in Fig. 3-1. The measuring technique is simple. First, the output impedance of the transceiver (52 Ω is most usual) is selected by one switch. Another switch is set to the type of measurement that you wish to make (SWR in the illustration). The middle slide switch on the unit in the figure is then set to FWD. Adjust the variable control for a full-scale reading on the meter, representing the peak standing-wave voltage or current. Next, set the switch to REF. When set to this position, the meter pointer indicates the ratio of the maximum to the minimum voltage or current on the transmission line, or the SWR.

SWR must be minimized before going through the procedure to

FIG. 3-1 SWR/power/field strength meter. (Courtesy Mura Corp.)

measure output power. After SWR has been adjusted to about 1:1, set the SWR/POWER switch to POWER. Use the 10-W range, as it will provide you with the most accurate reading on the WATTS scale. If SWR is 1:1, the meter pointer should indicate an output power, anywhere between 3 and 4 W. If power differs by much from the specified value, either you have an inaccurate instrument or your transmitter–antenna system is defective.

Although it is not absolutely essential, SWR is usually adjusted and power measurements made before checking the field strength. *Field strength* is an indication of the quantity of signal radiated by the antenna. The antenna must, of course, be connected to the antenna output jack on your transceiver when making these tests. (*Never* press the push-to-talk button on your microphone to transmit unless the antenna is connected to the CB radio. The connection may be direct or through an SWR or power meter.) A special pickup wire is connected to the field-strength meter to sense the radiated signal from the antenna. Settings of the slide switches on the panel of the tester are usually immaterial. The instrument is held a few feet from the antenna, and the variable control on the instrument is adjusted to the setting at which the pointer on the meter deflects to mid-scale. Do not stand between the antenna and the instrument, as this could materially affect the readings. It is also best to make the relative measurements while the instrument is on an insulated bench or table, as readings will differ from those made when holding the tester in your hand.

After going through all this, you may justifiably ask why. Actually, the field-strength measurement and reading are usually not very important to the performance and operation of the CB rig. The prime value is to determine the directional radiation characteristics of your antenna. From these measurements, you can determine if the field strength or radiated voltage from the antenna is stronger in one direction than in another. It probably is. If radiation is directional, then you should ask several questions. Is it due to some obstruction on the car or truck? Is it better to move the antenna to another spot so that its radiation pattern will not be as directional? Or, better still, do you need a dual antenna system for diversity reception? Or do you really prefer the directional characteristic?

When used properly and by applying some imagination, the field-strength meter can be a valuable accessory. After adjusting factors affecting the field strength to your liking, the instrument should be replaced in your setup to continuously monitor SWR and power.

Power Indicators

The wattmeter measures power when the antenna is the load. As discussed, measurements are accurate only if the antenna and transmission-line impedances are identical. For an accurate measurement of the capability of your transceiver to deliver power, a noninductive resistor load can be substituted for the antenna at the antenna terminal of the wattmeter. The resistor must be 50 or 52 Ω. A convenient package housing this resistor in an antenna connector plug is shown in Fig. 3-2. A second important advantage in using the resistor load is that it will not allow any substantial amount of signal to be radiated while making the power measurement. You will thus avoid interfering needlessly with other transmission while testing your rig.

Figure 3-3 shows a combination of antenna connector and load resistor bulb. This accessory can be connected to the antenna terminal on the transceiver instead of connecting the wattmeter and resistor load. The device will serve the same function as the test instrument when an exact reading is not essential, which is most of the time. RF power delivered by the transceiver causes the bulb to light. It glows brighter when there is more output from the CB radio than otherwise. Once you know

FIG. 3-2 Resistor load—52 ohms—in a PL-259 rf antenna plug. (Courtesy Gold Line Connector, Inc.)

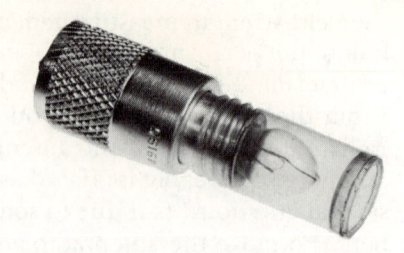

FIG. 3-3 52-ohm resistor load is light to indicate power output, mounted in a PL-259 rf antenna plug. (Courtesy Gold Line Connector, Inc.)

how bright the bulb glows when the transceiver delivers its rated power (from tests made when a known 4 W is being delivered), you can thereafter use the light intensity as a guide to indicate that the transceiver is operating properly. The bulb power indicator is an extremely convenient device to use when making power tests while on the road.

Antenna Matcher

The antenna matcher, shown in Fig. 3-4, is usually connected between the coaxial cable to the antenna and the SWR meter. Its primary function is to have the antenna system present the optimum load to the output circuit of the transceiver. As such, the matcher is very valuable. When the match is perfect, the SWR reading is at its ideal 1:1. Because SWR varies with weather conditions, the setting on the matcher should be readjusted as required.

The antenna matcher consists of circuits composed of two variable capacitors and an inductor. Capacitors are adjusted using the knobs on the panel while the SWR is being observed. The adjustment is proper when the SWR reading is at or close to 1:1. The matcher should be made a permanent part of the installation and readjusted each time the SWR reading drifts from its ideal.

FIG. 3-4 Antenna matcher. Connect between the antenna cable and transceiver rf output jack. (Courtesy Mura Corp.)

FIG. 3-5 Two position antenna selector switchbox. (Courtesy Gold Line Connector, Inc.)

Antenna Selector

An antenna mounted at one location on a truck or car may be more sensitive to receiving a particular CB signal than an antenna mounted at another location. In some installations, two antennas are mounted several feet apart on one vehicle. One method of determining which antenna is best in a specific situation is to quickly disconnect the first one from the rig while a signal is being received, and immediately substitute the second antenna. You then choose the antenna that provides the superior audible signal. On some transceivers, relative signal strength can be determined from S-meter readings.

It is, of course, inconvenient and unsafe for a driver to change an antenna connection while his vehicle is in motion. An antenna selector switch, such as shown in Fig. 3-5, is a valuable accessory to perform the transfer easily and rapidly. Not only is it convenient, but it is a lot safer to use the selector switch while driving than to change connectors. It is also easier to determine which antenna is performing better when switching rapidly from one to the other, than by using the more tedious method of physically changing connector plugs.

Also note the time factor. If you should change antennas by hand, you must ask the person communicating with you to talk for an extended period of time while you perform the experiment. The switch provides you with the same opportunity to select the more effective antenna without needlessly tying up the crowded CB band. Besides, you do not miss any part of the conversation while switching.

Two-Transceiver Couplers

Suppose that you have two transceivers and one antenna. A two-transceiver family is not a far-fetched idea. You may have one trans-

FIG. 3-6 Twin rig transceiver coupler. Use when two transceivers are to be connected to one antenna. (Courtesy Gold Line Connector, Inc.)

ceiver in the upstairs bedroom and one in the downstairs basement. You may even have two transceivers side by side to listen to one channel and talk on another, or to talk to two individuals at more or less the same time. You can accomplish this by merely flipping a switch on a coupler. With a little imagination, you can find many situations in which a second transceiver is useful and possibly even important to you. The coupler shown in Fig. 3-6 gives you the flexibility of using one antenna with two CB radios. It works in the following manner.

While the antenna cable is connected to the jack provided for it on the coupler, each transceiver is connected to one of the two output jacks.

The switch on the accessory has three positions. You can receive different (or the same) channels on both transceivers when the switch is set in the center position.

Only one CB radio can be used at any one time for transmitting purposes. Assume that you wish to transmit through the transceiver connected to output 1 on the coupler and to receive on the transceiver connected to output 2. Set the switch on the coupler to position 1 when transmitting and to position 2 when receiving. Although one antenna cannot serve both functions simultaneously, it can be switched from one transceiver to the other, so that you can alternate between transmitting on one unit and receiving on the second one.

Cophaser

Two antennas can be connected to a single transceiver at the same time. They may be mounted on opposite sides of a truck cab or car. The primary advantages of the arrangement are to minimize fading, increase the transmission range owing to improved antenna efficiency, and reduce the directional characteristics of the system.

In the transmit mode, the dual antenna system performs all its functions adequately only when the signals reaching the two antennas are in phase. That is, the peak in the RF sine wave (see Fig. 2-8) must be at a peak on both antennas at the same instant. Crests must likewise be present simultaneously on both radiating elements. Signals reaching the antennas can be phased properly through use of the cophaser shown in Fig. 3-7. First, adjust the individual antenna systems for minimum SWR. Next, connect both antennas and the antenna jack on the CB radio to the accessory. Adjust the control on the cophaser for the best SWR of the combination.

In the receive mode, the cophaser serves as a junction box for adding or combining the signals received at the two antennas and feeding the combination to the transceiver.

FIG. 3-7 Cophaser for connecting two antennas at the same time to the transceiver. Optimize output by setting control for minimum SWR. (Courtesy G.C. Electronics.)

Lightning Arrestor

Any object that protrudes into the air high above others is the most likely to be hit by lightning during an electrical storm. A base station antenna is a likely victim.

Electricity due to lightning at an antenna can be conducted through the transmission lines to the transceiver. Voltage (or energy) may be sufficient to destroy your valuable base station radio. The problem can be alleviated by supplying an alternate path for the electricity. Some protection is provided by grounding the antenna support. The lightning arrestor shown in Fig. 3-8 offers excellent additional protection.

The center conductor of the lightning arrestor is connected between the antenna and transceiver. Its function here is to complete the RF signal path. The outer shell of the arrestor is grounded, possibly to a water pipe or radiator. High voltage due to lightning will arc over from the signal conductor to the outside wall, and be diverted to ground rather than to the transceiver. When there is no voltage across the arrestor, it is an open circuit. The arrestor does not affect normal transmission or reception in any way.

FIG. 3-8 Lightning arrestor to protect your equipment. (Courtesy G.C. Electronics.)

Modulation Meter

Modulation was described with reference to Fig. 2-8. The amount of modulation is measured in terms of percent of modulation, which can be determined from the drawing in the figure. Multiply 100 by the difference between the voltage of the maximum positive modulation (peak) and the voltage of the maximum negative modulation (crest), as in Fig. 2-8c, and divide by the peak-to-peak RF carrier voltage in Fig. 2-8a.

More than 100 percent modulation is not permitted by the FCC. To prevent overmodulation, all new transceivers have limiting circuits. But for the maximum radiating power, modulation should be at a maximum (as close to 100 percent as possible). The modulation meter shown in Fig. 3-9 can be used to measure percent of modulation when connected between the transceiver antenna output and the antenna. A 52-Ω dummy load should be used instead of the actual antenna when making measurements.

To check modulation properly, first press the push-to-talk button on the microphone to place the transceiver into the transmit mode. Follow the procedures supplied with your modulation meter to be sure that all controls are set to accurately measure percent of modulation. Whistle (as loud as possible) into the microphone and read the percent of modulation on the meter. If you talk, the reading will not be constant, but will vacillate between one value and another, and the pointer on the meter will not indicate the true percent of modulation because it reacts too slowly. A steady tone created by whistling will give the meter pointer enough time to deflect fully and maintain a steady reading, accurately indicating the percent of modulation. Any reading above 80 percent is adequate when you consider the tolerances of components in the meter circuit, as well as the rated accuracy of the meter movement itself.

FIG. 3-9 Transceiver tester which includes facility to measure percent modulation. (Courtesy Mura Corp.)

Television-Interference Filters

After the expense of purchasing a CB rig, you may not feel like spending more money to avoid interference with your neighbors' television. But complaints from your neighbors may bring down the wrath of the FCC upon you. No one with an expensive rig wants that to happen, and so you purchase television-interference (TVI) filters to avoid annoying your neighbor.

Frequencies in the TV band can be generated in the transmitter section of your transceiver. Although they are not supplied by design, these frequencies are a result of harmonics or multiples of the fundamental 27 MHz generated in the CB radio.

A low-pass TVI filter, such as one of those shown in Fig. 3-10, should be installed between the antenna terminal on your transceiver and the antenna. It is intended to keep the harmonics from reaching the antenna. The filter allows only those frequencies that are below the TV band to reach your CB antenna while severely attenuating the higher frequencies. The conventional fixed TVI filter in Fig. 3-10a is usually sufficiently effective, but unusually difficult interference problems can be solved through the use of a tunable type of accessory. By employing a variable type of filter, such as shown in Fig. 3-10b, the rejection circuit can be more accurately adjusted to perform better in excluding the unwanted frequencies.

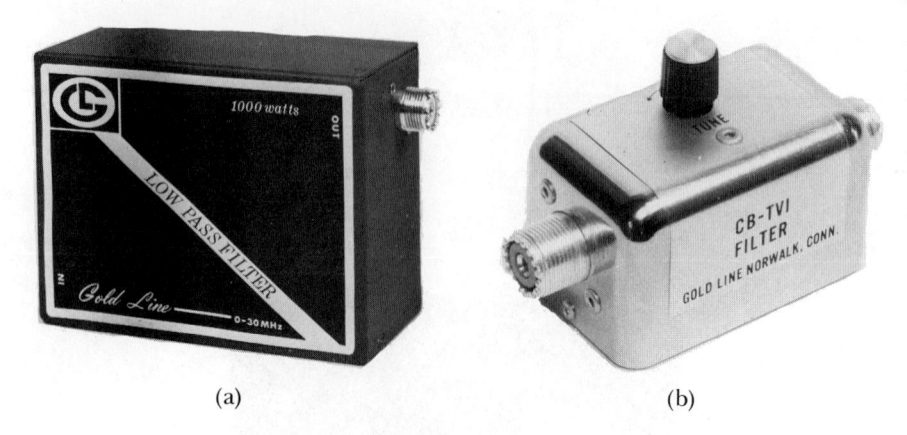

FIG. 3-10 Low pass TVI filter. The filter in (a) is fixed while the one in (b) is tunable. (Courtesy Gold Line Connector, Inc.)

Let us say that you do not eliminate all TVI through use of the filter at your antenna. A small gift to your neighbor, such as the high-pass filter shown in Fig. 3-11, should satisfy him by further alleviating the difficulties. This filter is connected between the TV antenna and the antenna terminals on the TV set. Because it is a high-pass filter, it attenuates frequencies below the TV band, such as those generated by your CB transceiver, while permitting signals at TV frequencies to pass freely.

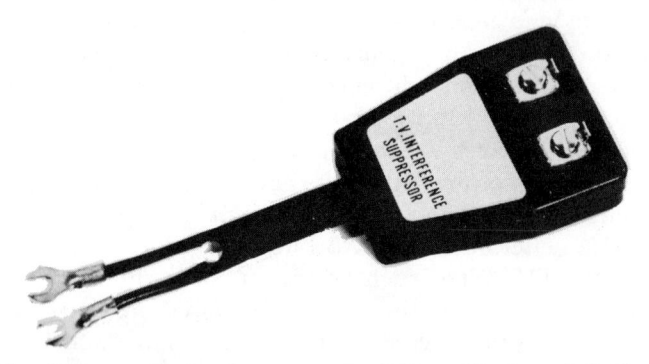

FIG. 3-11 High pass filter to be connected to TV set. (Courtesy G.C. Electronics.)

Power Supply

Base station CB radios are used at specific or fixed locations, such as in the home or operations center. The rigs are used primarily for communicating from the fixed location to a mobile unit. Transceivers made

Power Supply 45

FIG. 3-12 Filtered and regulated power supply for converting ac voltage at a wall outlet to dc power for a transceiver. (Courtesy Philmore Manufacturing Co., Inc.)

especially for this purpose are available, and have built-in circuitry so that they can be powered by the 120-V ac available at a wall outlet.

A base station may be established using a mobile CB radio that requires a nominal 13.8-V dc power supply. (It should operate properly with supply voltage ranging from 11 to 14.6 V.) A storage battery can be used, but it is bulky, dirty, dangerous, and heavy. Furthermore, it needs maintenance, such as adding water and acid from time to time, and recharging. A simple solution is to use a power converter, commonly referred to as a power supply, such as shown in Fig. 3-12.

A converter changes the ac line voltage to a dc voltage. For most mobile transceivers, a source of 11 to 14.5 V with a capability of delivering 2 amperes (A) is sufficient. Many supplies are available. Some are unregulated, so that the output voltage changes with the amount of amperage or current drawn from it at any one time. Although the change is undesirable, it can be tolerated if it does not provide more than 15 V when the transceiver is in the receiving mode and less than 11 V when transmitting. It is best that a regulated supply be used where the variation of output voltage is negligible when the line voltage changes from 105 to 132 V and when the load current varies from 0 to 2 A.

Ripple is an important characteristic of the power supply. Power is supplied to the power converter from the 120-V ac available at a wall outlet. Like RF, the voltage from an ac power source varies during each cycle. Because it is usually a 60-Hz power source, the cyclic variation is 60 times a second. Transformers in power supply circuits utilize this variation to convert voltage from the high 120 V at the wall outlet to the lower 13.8 V. Cyclic variation appears as 120-Hz (2 × 60 Hz) ripple (small pulses) on the direct current at the output of the supply. If the ripple is high enough, it can be heard while receiving signals on the transceiver.

Because it is usually worse when large amounts of current are drawn from the supply, bad ripple will be riding on the transmitted signal. Good supplies are sufficiently well filtered so that the ripple is negligible. At the 2-A rating, the peak ripple voltage present at the output of a satisfactory supply should be less than 0.01 V (10 mV).

In addition to being a convenient accessory when using a mobile transceiver as a base station, a power converter can serve well in other situations. You will find them extremely valuable, useful, and convenient devices when servicing mobile CB radios in the shop.

Remote Loudspeaker and Headphone

Remote loudspeakers and headphones serve several valuable functions. For one, they are useful in noisy areas. If ambient noise is high, you can hear better if sound from the loudspeaker is beamed directly at you. Loudspeakers in transceivers mounted in a vehicle usually face the floor. You hear sound from them only by reflection. A remote loudspeaker facing you, or a headphone, will allow you to more easily hear the received signals.

Remote loudspeakers are also useful if you are to monitor a particular channel at some distance from the transceiver. You simply adjust all controls on the CB radio for the proper reception of the channel, connect the plug at the end of the cable on the remote loudspeaker to a jack frequently provided for it on the transceiver, place the loudspeaker at a convenient location, and listen to all signals received on the channel in question. Of course, you should carefully adjust the squelch control so that weak and remote signals are blanked out.

Two remote speaker jacks are usually mounted on the rear apron of the transceiver chassis. One jack is used when the switch on the CB radio is set for PA applications. The second jack is active when CB radio signals are being received. The plug on the end of the cable connected to the remote loudspeaker is inserted into the proper jack, depending upon the setting of the CB–PA switch on the front panel of the radio. Circuitry is so arranged that the local loudspeaker is usually cut out when the plug for the remote loudspeaker is inserted into the appropriate jack. A circuit such as shown in Fig. 3-13 can be used.

The circuit from the audio output in the transceiver to the loudspeaker inside the CB radio is completed through leaves A and B of the jack. To operate the remote loudspeaker, its plug is inserted into the jack. The tip makes contact with leaf A, while the sleeve is connected to the chassis ground through the jack's outer structure ring and mounting hardware, C. Circuits from the remote loudspeaker to the audio signal in the transceiver through leaf A and ring C are now complete.

Remote Loudspeaker and Headphone

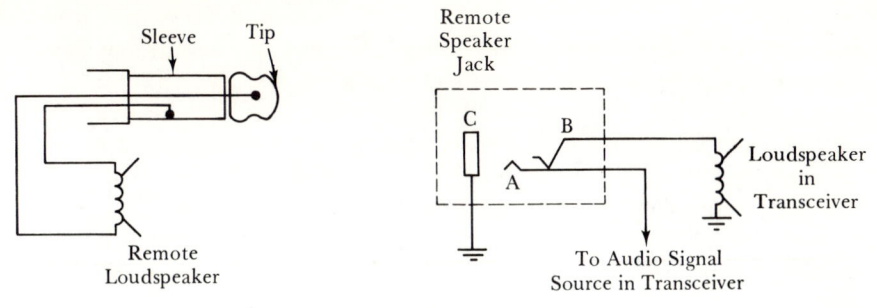

FIG. 3-13 Circuit for remote speaker jack.

Meantime, at the jack the tip pushed leaf A away from leaf B, disconnecting the loudspeaker in the transceiver from the audio output signal. Audio is now available only at the remote loudspeaker. Headphones could have been used rather than a remote loudspeaker so as not to disturb anyone else in the room.

Remote loudspeakers frequently provide better audio quality than those built into the CB radio. Even for normal operation, the remote speaker jack may be used as a convenient connector or outlet for a better loudspeaker. It may be placed right next to the transceiver.

As for the headset, it may be combined with a microphone into one structure. Known as a boom-headset, the microphone position is adjustable so that it can be located at an optimum position near the mouth. The plug at the end of the headphone cable is connected to the remote speaker jack, and the microphone connector is inserted as always into the microphone input jack (or plug) on the transceiver. A boom-headset is shown in Fig. 8-7. The lever on the box attached to the cable is used to switch the transceiver from the receive to the transmit mode. The knob is for adjusting the listening level.

4

Tests and Test Equipment

The heyday of the screwdriver mechanic has long since passed into oblivion. It is the unusual serviceman that still makes house calls armed only with a screwdriver as his test gear. Well-equipped service centers enable the technician to analyze problems arising in electronic equipment, correct the defects, and then use sophisticated instruments to assure himself that the once-defective item is now operating at its optimum capability.

Few pieces of equipment need be added to that already in most shops to enable the serviceman to handle CB radios properly. Some of the instruments may be relatively expensive. Considering the tremendous proliferation of CB rigs, a one-time investment in good test instruments can prove to be financially rewarding within a very short period of time. On the other hand, don't go overboard when buying test equipment. Remember that all you want to do is install, troubleshoot, and repair CB radios. Do not buy instruments useful only in design situations. Instruments of this type will just sit on your bench gathering dust, and will prove to be a distinct waste of money.

Test instruments are frequently divided into two groups: (1) testers used on the bench and (2) instruments used primarily in the field. Some equipment fits into both groups. Instead of making artificial categorizations for the various pieces of equipment, special applications will be noted as some of the individual types of instruments are described. Circuits and components found in CB radios are discussed where applicable with but one very important consideration in mind. That is to supply enough details about the transceiver to enable you to choose properly the least expensive instrument to do the entire job, without buying unnecessary features. The discussions can also serve as a review of circuits and devices that must be tested with insight if the overall servicing operation is to be financially rewarding.

Measuring Voltage and Resistance

Unless the technician is still doing everything by screwdriver, the service center he is working in must have some type of instrument to measure voltage and resistance. Different types of instruments are now being used. More than one volt–ohm–milliammeter (VOM) can normally be found on the test bench. Vacuum-tube voltmeters (VTVMs) serve the same functions as the VOM, but because VTVMs also embody electronic circuits, these instruments present a higher impedance to the circuit being measured than do VOMs. High impedance is, of course, important if voltages in the circuit being tested are not to be affected by the presence of the test instrument while the measurements are being made.

VTVMs have been or are being rapidly supplanted by transistor voltmeters (TVOMs) using transistors rather than vacuum tubes in the electronic circuits. Finally, there is the digital volt–ohm–milliammeter (DVOM), which can have advantages over the more conventional VTVM and TVOM, which use meter movements as the indicating devices.

Many DVOMs are more accurate than other types of measuring instruments and are thus excellent pieces of laboratory equipment. Although DVOMs can serve well in the shop, they have the disadvantage that you cannot visualize rapid voltage changes as readily on instruments using digital readouts as on instruments using meter movements. Inexpensive DVOMs are usually no more accurate than ordinary TVOMs, although DVOMs are easier to read. The response time of the cheaper units is frequently very slow. If you like DVOMs, by all means continue using them. Our prejudice for service work is toward the "old-fashioned" and less expensive test instrument using a meter movement as the indicating device. An instrument that uses both—the digital readout and meter movement—as indicators is truly ideal.

Measuring Voltage and Resistance 51

FIG. 4-1 Mura model 90-M—50,000 Ω/V dc VOM. (Courtesy Mura Corp.)

Multimeters that can be used for troubleshooting CB equipment in the service shop are shown in Figs. 4-1 and 4-2; a convenient portable instrument for use in the field in installing and servicing applications is shown in Fig. 4-3.

Resistances of the dc voltage ranges of the Mura 90-M VOM in Fig. 4-1 are high. Each range is specified as having a ratio of resistance to full-scale voltage or "sensitivity" of 50,000 Ω/V. To determine the resistance of each dc voltage range, multiply the maximum voltage that can be applied to that range by 50,000. For example, the resistance of the meter when the switch is set to the 25-V range is 25 × 50,000 or 1,250,000 Ω. This is quite high and usually more than satisfactory when the instrument is applied to troubleshooting a CB transceiver. It is unlikely that a resistance of this magnitude will load a circuit sufficiently to affect the readings. The VOM features a carrying handle so that it can be taken to a job in the field, should an instrument of this caliber be required.

Even higher resistance on dc ranges can be realized by using the Eico 242 TVOM in Fig. 4-2. Resistance on all dc ranges is 11 MΩ. For an electronic multimeter, the instrument in the figure is unusually versatile; it can be used to electronically measure dc and ac current, as well as the factors that can be measured on all other TVOMs, such as dc and ac

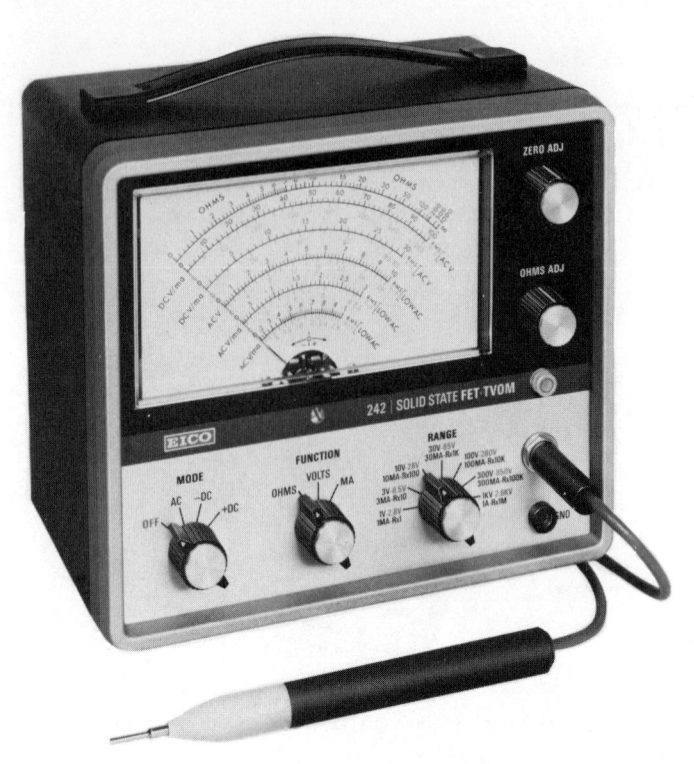

FIG. 4-2 EICO model 242 TVOM. (Courtesy EICO Electronic Instrument Co., Inc.)

voltage and resistance. Because it uses a large meter movement, it is easy to read and suitable for use on the bench.

One precaution must be observed, especially when using VTVMs and TVOMs near strong RF fields. The pointer on the meter may deflect owing to the effect of the RF field on the electronic circuit, even when the meter is not actually connected to the CB radio. Before making measurements using an electronic multimeter, connect one test lead to the other. Establish an RF field by pushing the press-to-talk button on your microphone. Note if the meter pointer deflects from zero. If it does, move the instrument to a location on your bench where the field will not produce this effect.

The Mura NH-55 multimeter in Fig. 4-3 is a relatively low-resistance VOM. Although its 2000-Ω/V characteristic is adequate for troubleshooting most circuits in CB radios, it may provide too much of a load in a few isolated instances. Voltages in the circuit may change somewhat while the measurements are being made owing to the presence of the meter. Because of its size, the VOM is ideal for use in the field. Use it to check the supply voltage available for the transceiver, to check if the

Power Supply

FIG. 4-3   Mura model NH-55—2000 Ω/V dc VOM. (Courtesy Mura Corp.)

antenna system is in good order and has been installed properly, to check if a fuse is still good, and in many other situations.

At least one multimeter in each category should be available to the CB serviceman. All types discussed are inexpensive, and the purchase of all three will not require a big investment. A useful addition to the TVOM at very little cost is the RF probe, such as the Eico unit in Fig. 4-4. TVOMs can be used to measure RF voltage (using the dc voltage ranges) when it is applied to the instrument through the probe.

FIG. 4-4   RF probe. (Courtesy EICO Electronic Instrument Co., Inc.)

Power Supply

In Chapter 3, the Philmore PS-123 regulated power supply was described as an ideal accessory in many applications. The unit can readily and successfully be used by the technician in the service shop. But the

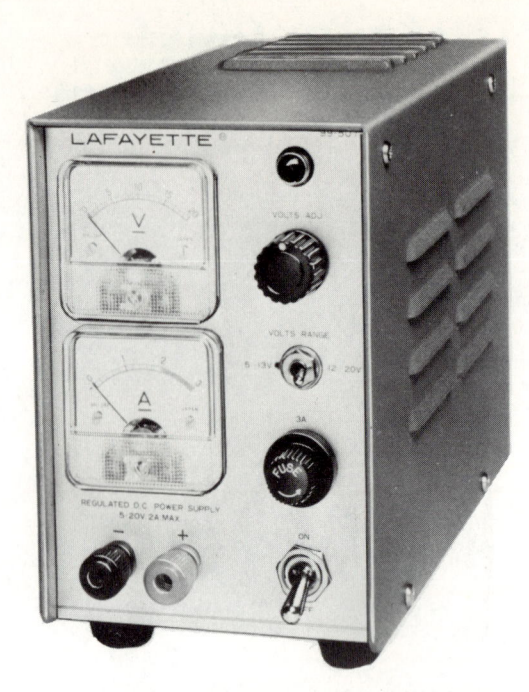

FIG. 4-5 Variable power supply. (Courtesy Lafayette Radio Electronics Corp.)

supply is limited to providing one fixed voltage. Every transceiver should operate properly at the 13.8 V that the supply provides.

Regardless of whether the supply voltage is as low as 11 V or as high as 14.4 V, the CB radio should perform within reason. A knob on the front of the Lafayette 99R50775 variable power supply shown in Fig. 4-5, can be used to set the output from the instrument to well below the 11 V and to considerably above the 14.4 V considered the supply voltage extremes for proper transceiver operation. After its output has been set, the regulated voltage at the terminals and the current drawn from the supply can be determined from readings on the two meters mounted on the supply's panel.

A variable supply voltage is extremely useful when troubleshooting. High and low voltages should be available for the transceiver so that you can check its performance at all supply voltages that may be encountered in the field. A high voltage is also useful when you are looking for an elusive problem component. It may behave properly (or erratically) when the transceiver is supplied with its normal voltage, but may break down when the voltage is raised above the average. Applying a momentary surge of voltage somewhat above normal to the CB radio may be sufficient to precipitate the complete breakdown of the intermittent component. It is infinitely easier to locate a defective part in the CB radio than to hunt for an intermittent item. Continuous operation at somewhat above 14.4 V should not damage the transceiver. It can usually be powered at the elevated voltage for a considerable amount of time. Time should be sufficient for you to locate the defective component and yet not result in damage to any sound item in the CB radio.

Transistor Testers

Most components in transceivers can be checked on one of the various types of multimeters already described. It is obvious that the ohmmeter ranges are used to measure resistors. The highest ohms range can also be used to determine if a capacitor is good or bad. Any movement of the pointer on the meter when the capacitor under test is first connected to the test leads is a fair indication that the capacitor is not open. After its initial movement toward the low-resistance end of the scale, the pointer will come to rest at a point on the scale indicating a higher resistance. An infinite resistance reading when testing ceramic and mylar capacitors and a very high resistance indication of well over 500,000 Ω when checking electrolytic and tantalum capacitors indicate the device under test is good. Low-voltage ceramic capacitors are good even if the final reading on the meter if only somewhat above 1 MΩ.

Solid-state devices can also be checked on multimeters by noting the ratio of the forward to reverse resistances of each junction. Even though this test does not provide conclusive proof that the semiconductor device is good, chances are that, if it passes the ohmmeter test, the device is in proper operating condition. Absolute quality checks of solid-state devices can only be made using a transistor tester or, better yet, a curve tracer. As good curve tracers are very expensive, we recommend that each service shop be equipped with at least one good transistor tester.

Because transistors are usually soldered into printed circuit boards, it is inconvenient to remove the devices from the board to determine if they are good. Besides, any excess heat applied when unsoldering a transistor from the board may damage the device. It is thus recommended that an in-circuit tester be used.

There are a number of different types of in-circuit testers on the market. Before proceeding with a discussion and comparison of the available types of equipment, it would be useful to review some of the theory of solid-state devices and to present some procedure for troubleshooting the circuits around these devices.

Semiconductor Materials

Practically every modern semiconductor component is composed of either germanium or silicon slabs diluted or doped by another material. (Copper oxide rectifiers are still being used in the ac voltage section of VOMs by some manufacturers, but their use is becoming less frequent.) In the pure state, germanium and silicon are fair insulators. Once doped, conductivity increases somewhat, and they become what are known as semiconductors. The ability of semiconductors to conduct electricity lies somewhere between that of insulators and conductors such as copper.

There are two types of semiconductor materials. One has negatively charged electrical particles, or electrons, loose in its structure. The other has spaces in the material because of a shortage of free electrons. These spaces are referred to as holes. Accordingly, a semiconductor slab with free electrons is referred to as an *n*-type material. Its counterpart, the *p*-type material, lacks free electrons.

Junction Diodes

Junction diodes and power rectifiers are formed by joining an *n*-type slab with a *p*-type slab, as shown in Fig. 4-6a. The symbol for the diode is shown in Fig. 4-6b. One lead is connected to the *p* slab and is referred to as the anode. The remaining lead is connected to the *n* slab, the cathode. The diode conducts current, *I*, freely when the anode is positive with respect to the cathode, and resists the flow of current when voltage is

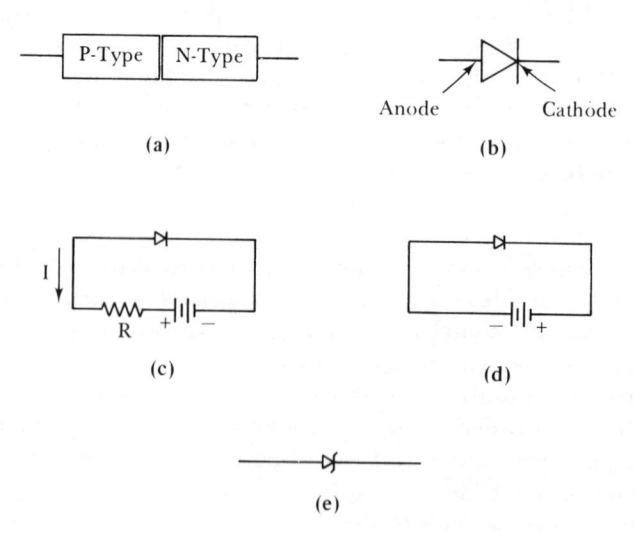

FIG. 4-6 Junction diode.

Transistor Testers

applied to the device in the opposite direction. This is shown in Figs. 4-6c and 4-6d, respectively. When the anode end is the more negative of the two, only a slight amount of leakage current does flow. It is best when leakage current is at a minimum. One reason for the predominance of devices composed of silicon material over those made of germanium is that leakage current is much lower in the diode and transistor made silicon semiconductors. The resistor R shown in the circuit in Fig. 4-6c is to limit current to safe values for the diode, so that is will not be destroyed by an excess amount of current flowing through it.

As the potential across the reverse-biased diode is increased, a specific voltage is reached at which the diode "breaks down" and conducts a significant amount of reverse current. A resistor must also be placed in series with the diode to limit this reverse current to safe values. Voltage across any resistor or circuit placed across the diode is maintained at the reverse breakdown or zener voltage of the diode. (This follows from the obvious rule that voltages across all components connected in parallel are equal.) Zener diodes are special devices selected for their specific breakdown voltage. Voltages across zener diodes are maintained relatively constant regardless of the current flowing through them, and are thus useful as voltage regulators. The symbol for the zener diode is in Fig. 4-6e. It is connected in the circuit as in Fig. 4-6d, but with an added resistor in series to limit reverse current to values that the diode can pass safely.

Voltage across the forward-biased diode in Fig. 4-6c is also relatively constant when current flows through it. The voltage across a germanium device is anywhere between 0.2 and 0.4 V; across a silicon diode it is 0.6 to 0.8 V.

When connected in a circuit, diodes can be checked with a voltmeter. If current flows through the diode, voltage across a good device must be as just indicated for the germanium and silicon diodes. When used as a zener, the voltage across the reverse-biased diode must be at its rated breakdown voltage.

An ohmmeter can be used to test any type of diode when at least one of its leads is disconnected from the circuit. Before making tests with an ohmmeter, we must determine which of the meter leads is positive with respect to the other lead. The red or black lead from the instrument may be the positive lead. A simple procedure to determine polarity requires the use of a second dc voltmeter. Connect the leads from the ohmmeter to those on the voltmeter in the proper way for the voltmeter pointer to move up scale. When this happens, the positive lead of the voltmeter is connected to the positive lead of the ohmmeter.

Ohmmeter current must be limited to 3 milliamperes (mA) if the diode is to be tested safely. Use the following procedure to determine the current from your ohmmeter. First, read the voltage at the ohmmeter leads using one of the ranges on the second voltmeter. Then note the resistance that can be read at the center of the ohmmeter scale on the particular ohmmeter range that you are using. The voltage at the test leads, divided by the resistance at the center of the scale, must be less that 0.003, if that ohmmeter range can be used to safely test a diode or transistor. If the quotient is greater than 0.003 (indicating that the ohmmeter can supply more than 3 mA in this particular range), use the next higher resistance range. Repeat the measurements and calculations until you find a range where the quotient is less than 0.003. Use the lowest range that can be applied safely.

Once the polarity of the leads is known, and the lowest safe resistance range has been determined, connect the diode to the ohmmeter. The meter should indicate a low resistance when the anode is connected to the more positive lead of the ohmmeter and a high resistance when the connections to the leads are reversed.

Bipolar Transistors

Bipolar transistors can be thought of as consisting of two junction diodes joined together. Transistors are shown in Fig. 4-7a; their diode equivalents and transistor symbols are shown in Figs. 4-7b and 4-7c, respectively.

Two types of transistors are shown, *npn* and *pnp*. The type of transistor formed from the semiconductor material depends upon the types of slabs comprising the device. Obviously, if the base is of *p*-type material, the collector and emitter must be *n*-type slabs and the final device is an *npn* transistor. Throughout most of the discussion, *npn* transistors will be used as examples; *pnp* types behave in nearly identical ways. To apply the discussion on *npn* transistors to *pnp* devices, just reverse the polarity of voltages indicated for the *npn* transistors. As an example, if the base is positive with respect to the emitter in a circuit involving an *npn* transistor, the base must be negative with respect to the emitter for the identical circuit when *pnp* devices are involved.

Considering the *npn* transistor circuit in Fig. 4-8, the collector must be positive with respect to the base and emitter if electron current is to flow from the emitter to the collector. All transistor current flows through the emitter. As it leaves the emitter, the current is divided between the collector and base. Almost all of it goes to the collector. The ratio of the collector to emitter current, I_C/I_E, is the alpha of the transistor, denoted by the symbol α. Numerically, it is just slightly less than 1.

The symbol β (beta) denotes the ratio of the collector to the base current, I_C/I_B, or the dc current gain of the transistor from the base to the collector. Beta may range from less than 10 to over 1000. In many

Transistor Testers 59

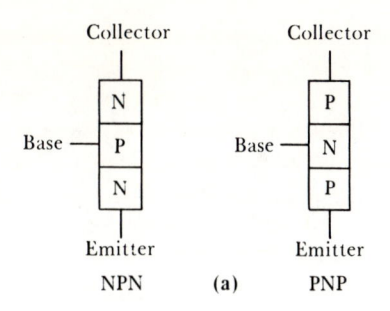

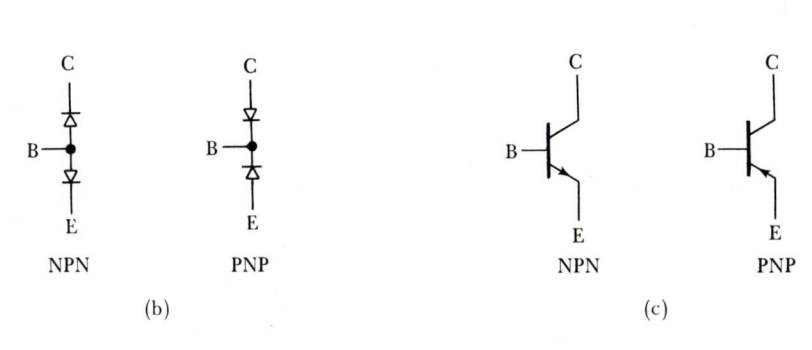

FIG. 4-7 Junction transistor.

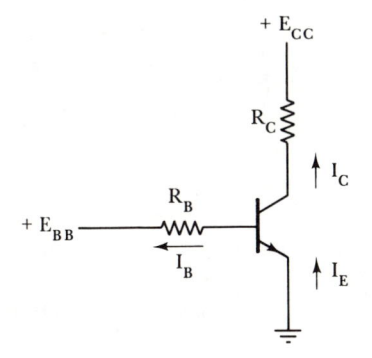

FIG. 4-8 NPN transistor circuit.

applications, beta may be considered as a fixed number, which does not vary to any degree with variations in collector current or collector to emitter voltage.

The amplitude of ac current (RF or audio) varies through the cycle. Apply this ac current change to the base of a transistor. Assuming beta as a constant, the collector current will vary in step with the ac current at the base. Collector current changes are greater than base current changes. Because a small current change in the base produces a large current change in the collector, the junction transistor can provide ac

(a) Common Emitter

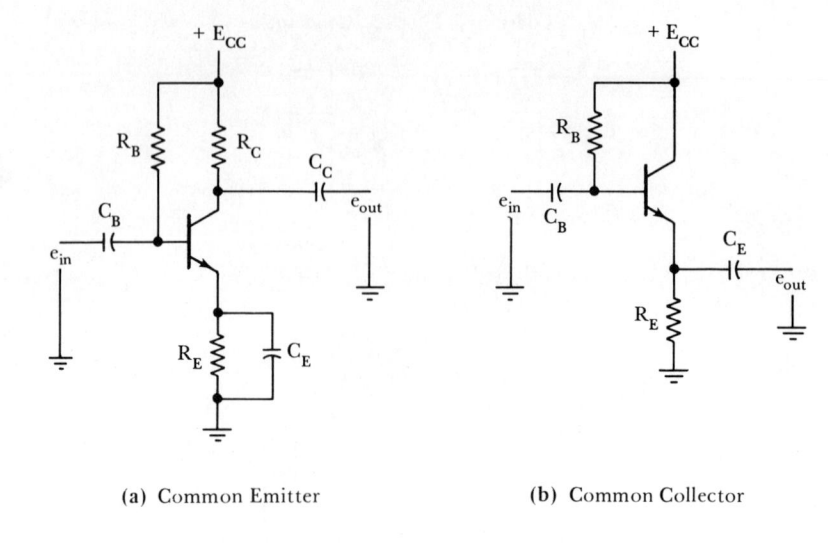

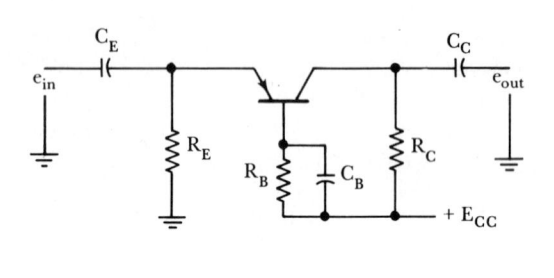

(b) Common Collector

(c) Common Base

FIG. 4-9   Basic transistor circuits.

current gain. The ratio of the collector to base current change is the ac beta of the transistor.

Put a load, regardless of whether it is a resistor as R_C or an inductor, into the collector circuit, as in Fig. 4-8. Voltage is developed across the load owing to the collector current flowing through it. As voltage is higher across R_C than at the base, a transistor in this circuit can also provide voltage gain.

We diverge here for a few paragraphs from the basic discussion of transistors to indicate the three basic circuit configurations used in CB radios. They are shown in Fig. 4-9. More specific circuits will be discussed in later chapters.

When considering the dc operation of a circuit, the battery or B+, shown here as $+E_{CC}$, is E_{CC} volts above ground. This is only true of the

Transistor Testers

dc characteristic. As far as the alternating current and RF are concerned, the battery is *all* ground. For signal purposes, the battery is an ac short, and E_{CC} is at ground potential. This state is frequently assured through the use of a capacitor connecting the positive or + terminal of a supply to the negative or grounded − terminal. Filter capacitors in dc power supplies are used to bypass alternating current (or ripple) from B+ to ground.

The most commonly used transistor circuit, the common-emitter arrangement, is shown in Fig. 4-9a. The base–emitter junction is forward biased from the $+E_{CC}$ supply through R_B and R_E. Signal is applied between the base and emitter at e_{in} as C_E is a short circuit from the emitter to ground at signal frequencies. The output is across R_C, between the collector and emitter. It is obvious that e_{out} is across R_C as well as the collector and emitter when it is recalled that E_{CC} is at signal ground and that C_E is a short circuit to ac across R_E, clamping the emitter to signal ground. If C_E were not in the circuit, R_E would be an element common to both the collector and emitter circuits. In this capacity, a feedback voltage would be developed across it, with a consequent reduction in voltage gain. The voltage gain of a common-emitter circuit is generally high, current gain is about equal to the beta of the transistor, and the input and output impedances are usually several thousand ohms.

In the common-collector circuit shown in Fig. 4-9b, all signal voltages are considered with respect to the collector. The collector is at ac ground. The input signal is applied between the base and collector; the output is between the emitter and collector, across R_E. The ac input impedance of this type of circuit is normally high while the output impedance is very low. The circuit thus lends itself ideally for use in isolating the input from the output. It is also referred to as an emitter follower. The voltage gain is slightly less than 1; the current gain is the beta of the transistor.

The common-base circuit shown in Fig. 4-9c is useful as an RF amplifier. Recalling that $+E_{CC}$ is at ac ground and that C_B is an ac short circuit, e_{in} is applied between the emitter and base, and e_{out} is between the collector and base. The input impedance of this type of circuit is low; the output impedance is the same order of magnitude as is the output impedance of the common-emitter circuit, usually equal to R_C. Although voltage gain can be considerable, current gain is slightly less than 1 and is equal to the alpha of the transistor.

Resistor input and output loads are shown. In RF circuits, loads are frequently inductors, such as IF transformer windings, chokes, and RF coils. Regardless of the type of load, the circuits described here are used in one form or another when bipolar transistors are in CB radios.

In any of the circuits, a quantity of base–emitter current controls

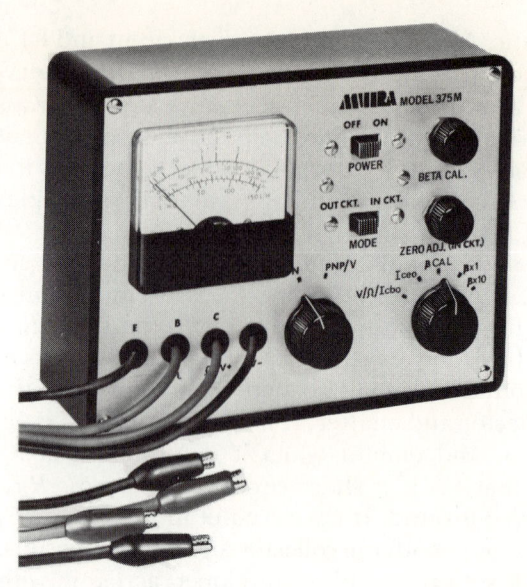

FIG. 4-10 Transistor tester. (Courtesy Mura Corp.)

the amount of collector current that flows. A fixed voltage is usually placed in series with the base and base resistor to set the average value of base (and hence collector) current. This is the bias current. In Fig. 4-8, E_{BB} is in series with R_B. Voltage across the base–emitter junction when the transistor is conducting is the same as that of a forward-biased diode, about 0.7 V for a silicon device and 0.3 V for a germanium transistor. These voltages are almost always present when the transistor is operating. The absence of the proper base–emitter voltage is often a pretty good indication of a defective solid-state device, but there are exceptions to the rule. More of this will be discussed as specific CB circuits are described.

Direct current should not flow from the collector to the base or from the collector to the emitter, because the base–collector junction is reverse biased. However, some undesirable leakage current always exists. The collector–base leakage current has been assigned the symbol I_{CBO}; the collector–emitter leakage current is I_{CEO}.

A bipolar transistor *circuit* can best be checked by noting the voltage on the various elements of the device. Testers are specifically designed to test the *isolated* transistor. More advanced types of testers can be used to check the transistor in the circuit, without being forced to first remove it from the printed circuit board. We should be involved only with testers providing this capability. Three testers that can be used to make in-circuit tests are shown in Figs. 4-10, 4-11, and 4-12. Each is designed to

Transistor Testers 63

FIG. 4-11 Transistor tester. (Courtesy EICO Electronic Instrument Co., Inc.)

perform different tests in different ways, but the final goals of all are the same—to indicate if the transistor is good or bad.

The most economical tester, the Mura 375M, is shown in Fig. 4-10. It can be used to measure dc beta regardless of whether the transistor is in or out of circuit. A control on the front panel can be adjusted to compensate for the effect on the beta measurement of components in the transistor circuit. If the transistor must be removed from the circuit to be absolutely certain that it is good (and this may be necessary when using any in-circuit tester if the resistances of components in the circuit are very low), the Mura 375M can be used to more accurately measure I_{CBO}, I_{CEO}, and dc beta. Facilities are also provided for making a simulated ac beta test. To add to the serviceman's convenience, dc voltage and resistance measuring capabilities are built into this instrument. Besides checking transistors found in CB transceivers, this tester can be used to determine if diodes are good, shorted, or open.

The Eico 685 shown in Fig. 4-11 is designed to measure ac beta in and out of the circuit. After the transistor has been removed from the circuit, ac beta can be measured accurately while the transistor is an independent device.

The beta of the various types of transistors can be measured most accurately only when specific amounts of collector current are present. It

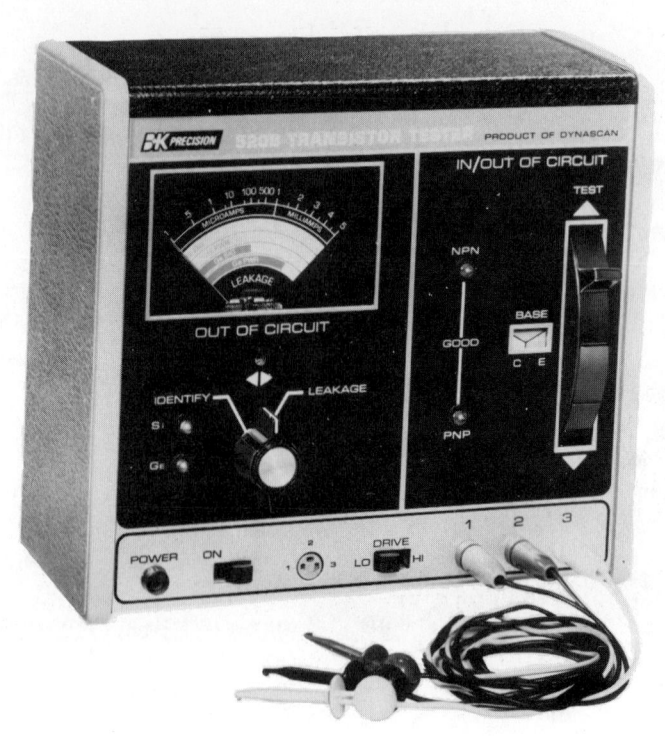

FIG. 4-12 Transistor tester. (Courtesy B&K Dynascan Corp.)

is recommended by the manufacturer of the Eico 685 that the beta of RF amplifying devices be measured when the collector current through the device is 0.2 mA. For small audio amplifier transistors, the measurement should be made when the collector current is 2 mA. Using similar logic, a power device requires 20 mA in the collector circuit if its beta is to be determined accurately. The Eico 685 ac transistor tester provides facilities to measure beta at all three preferred levels for the different types of devices; most other testers provide only one collector current level at which beta can be measured. Although this accuracy is nice to have, its absolute necessity in our application is a matter of divergent opinion.

The Eico 685 transistor tester can also be used to measure leakage current of transistors and diodes, as well as circuit voltage and resistance. It features a source of 20-V dc essential to determine the breakdown voltage of zener diodes as well as to indicate if the various junctions of the transistors and signal diodes can withstand voltages normally found in CB transceivers.

The design of the B & K Dynapeak 520 tester shown in Fig. 4-12 differs radically from the designs of the two instruments described above. The three leads from the tester are connected to the three leads from the transistor. This is true of all testers. But, unlike the others, with this tester you do not have to know which lead on the transistor is the collector, which the base, and which the emitter. You can connect any of the leads from the meter to any of the leads on the transistor. You do not even have to know if the transistor is a *pnp* or an *npn* device. Just connect the leads from the instrument to leads on the transistor. From tests made on this instrument, you can determine the type of transistor that is being tested, identify the base lead, and be advised if the solid-state device is made of silicon or germanium semiconductor material. All this can be done rapidly and regardless of whether the transistor is in or out of the circuit. Besides indicating to you directly if the device is good or bad, you can also use the B & K Model 520 to measure leakage current.

Which transistor tester should you get? You already have a considerable amount of information about all of them. You must, of course, make the decision for yourself. Perhaps you can make a better decision after the following brief review of field-effect transistors as applied to CB radios, and a discussion of how these transistors are checked on the three transistor testers.

Field-Effect Transistors

Field-effect transistors (FETs) are being used in many new transceivers. They have two very important advantages over their bipolar counterparts. For one, the input impedance of an FET is very high, because the input junction of an FET is reverse biased, whereas the base–emitter input junction of the bipolar device is forward biased. The second big advantage is the isolation of the input from the output. This characteristic is utilized in mixer circuits where it is desirable to isolate the local oscillator from the incoming RF signal.

Two basic groups of FETs are used in CB radios: (1) the junction FET (JFET) (2) the insulated-gate FET (IGFET), also known as a metal oxide semiconductor field-effect transistor (MOSFET or MOSTFET). JFETs are used more often than IGFETs, although both types are finding their way into the newer transceivers.

The JFET consists basically of a slab of semiconductor material to form what is known as the channel. If the slab is made of n-type material, the transistor embodying it will be referred to as an n-channel device. Obviously, p-channel transistor refers to a JFET using p-type material for the channel. The discussion to follow will be on n-channel JFETs. To apply the information to p-channel transistors, only the directions of the current flow and the polarity of applied voltages must be reversed. The same convention will apply to the discussion of IGFETs.

A lead is connected to each end of the channel, as shown in Fig.

4-13. One lead is referred to as the *source*; the other is the *drain*. A *p*-slab forms a junction with the channel, and is referred to as the *gate*. In conventional circuits, the *p*-slab is made negative with respect to the source, so the junction is reverse biased. This bias voltage affects the channel current, as the drain current is reduced when the reverse-bias voltage is increased. The bias voltage can be made so negative that less than 50µA of drain current flows. This is known as the pinch-off voltage, V_P, and is a very important characteristic of the transistor.

There can be leakage from the positive drain to the negative gate. The leakage current is normally very low. It is assigned the symbol I_{GSS}.

Another important JFET characteristic is the I_{DSS}. This is the channel current that flows when the gate to source voltage is made zero. This is almost the maximum source to drain current that a transistor can conduct and still amplify.

Amplification is the function of transistors. As was indicated, the amount of drain current that flows is controlled by the size of the gate to source voltage. Amplification is a ratio of how much the drain current, I_D, changes with a change of gate to source voltage, V_{GS}. The ratio, I_D/V_{GS}, the transconductance, has been assigned the symbol g_m, an extremely important JFET characteristic.

Cursory JFET tests can be made with an ohmmeter. For example, the channel resistance in either direction can be measured as being anywhere between several hundred and several thousand ohms on a good transistor. The gate–source junction behaves as does the junction of any other diode and can be checked accordingly. Other factors can only be measured using transistor testers.

A typical circuit using a JFET is shown in Fig. 4-13c. Signal voltage is applied between the gate and source; output voltage is developed across R_D in the drain circuit. Voltage across R_S is in the proper direction to make the source positive with respect to the gate or, stated more conventionally, to make the gate negative with respect to the source. The junction is thus reverse biased. If a capacitor is across R_S, the source is at ac ground. Otherwise, output signal voltage developed across the source resistor is fed back into the gate circuit, reducing the gain of the overall amplifier.

The behavior of the IGFET is very similar to that of the JFET, except that all junctions are isolated from each other. Schematic representations are shown in Fig. 4-14. Proper voltages are applied to the elements with respect to the source, as was done for the JFET, with the addition that the lead from the substrate is frequently connected directly to the source.

Gates of IGFETs are insulated from the other elements. ("Gates" is

Transistor Testers

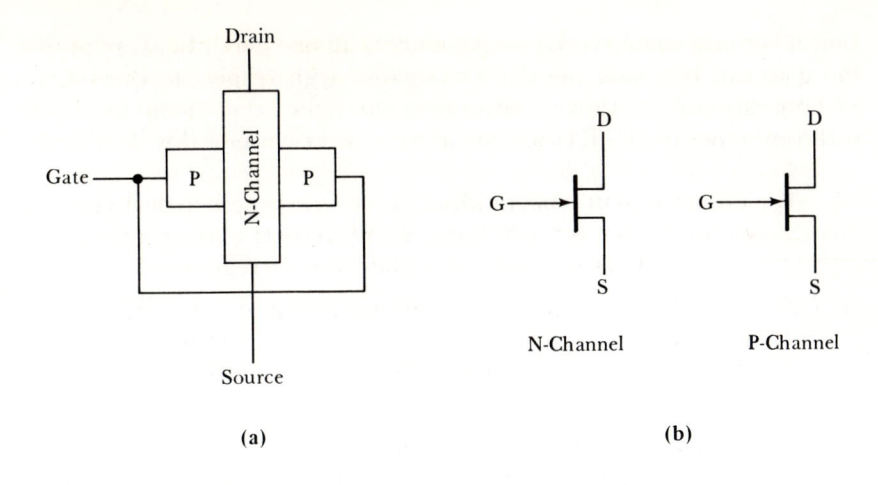

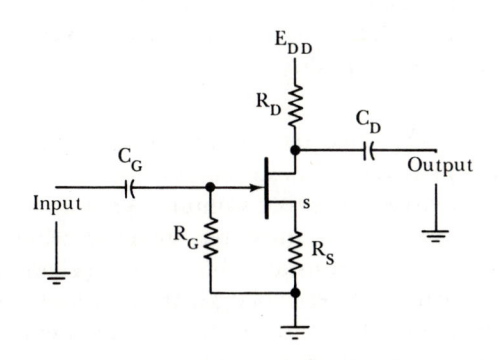

(a)

(b)

(c) Typical JFET Circuit

FIG. 4-13 The JFET.

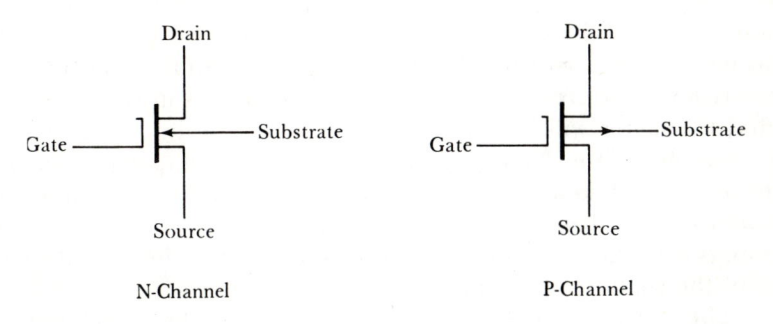

FIG. 4-14 Insulated Gate FET.

plural because some IGFETs have more than one gate.) Because of this, the gate can be made positive or negative with respect to the source, without any current flowing from the gate to the other elements. Three different types of IGFETs are in current use because of this flexibility:

1. Enhancement type, in which there is no source to drain current until the gate is made positive with respect to the source. More current flows when the bias voltage is increased.

2. Depletion type, in which drain current is at a maximum when the gate to source voltage is zero. Drain current is reduced as this bias voltage is made more negative. The JFET also belongs in this category.

3. Depletion and enhancement type combinations are quite common. There is drain current when the bias is at 0 V. Drain current decreases as the gate is made more negative with respect to the source, whereas current increases as the bias voltage is made more positive.

IGFETs can be classified by the same characteristics (such as g_m, I_{GSS}, etc.) as can the JFETs. These factors can only be measured on transistor testers (or curve tracers), as ohmmeter checks are not conclusive. Furthermore, precaution must be observed when handling an IGFET. Because of the high resistance between the elements, static electric voltage can be formed. These voltages are very high, and the insulation in the transistor may break down because of them. To avoid this type of catastrophic failure, all transistor leads are normally connected to each other until the IGFET has either been wired into its circuit or connected to a transistor tester.

The Mura transistor tester shown in Fig. 4-10 can be used to test several characteristics of FETs. Although not a quantitative test, the channel of a JFET can be checked to determine if it is in good shape. This test can be performed by connecting the ohmmeter on the instrument from the source to the drain. If the channel is in proper working order, the meter will indicate a reading of several hundred to several thousand ohms. Resistance readings can also be used to determine if the gate to channel junction of the JFET is good, using the procedures previously described for testing a junction diode. Forward resistance readings here should be similar to those found when checking the condition of the channel, if the junction is intact.

The dc transistor tester can, of course, be used to measure gate to channel leakage, I_{GSS}, for any FET, just as it can be used to measure collector to base leakage, I_{CBO}, of the bipolar device. Furthermore, I_{DSS} can be determined accurately on depletion-type FETs if the current is less than 5 mA and the pinch-off voltage of the device is less than 1.5 V.

The Eico tester shown in Fig. 4-11 can be used to test both JFETs and IGFETs, in or out of circuit. I_{DSS} can be measured to 50 mA on depletion-type transistors. In addition, there are facilities to measure I_{GSS}, g_m, and pinch-off voltage, V_P. The transconductance (g_m) can be measured with the transistor as an independent device, or when it is wired in the circuit.

FETs can be tested most rapidly on the B & K 520 tester shown in Fig. 4-12. Connect the three test leads from the instrument to the three leads on the transistor. You do not have to know beforehand which of the transistor leads is the source, drain, or gate. An indicator on the tester will help you to identify the gate lead. In- or out-of-circuit readings reveal whether the transistor is good or bad. Leakage, or I_{GSS}, as well as I_{DSS}, can be measured after the transistor has been removed from its printed circuit board.

All three testers will do the job you need. Which one to buy is up to you. If price is the prime consideration, the instrument shown in Fig. 4-10 is for you. Direct-current beta can be measured on this instrument; ac beta at three different collector current levels can be determined from readings on the tester shown in Fig. 4-11. In addition, determination of the ac transconductance of FETs can be made through the use of the latter instrument. After the device has been tested out of circuit, the beta or g_m readings should be compared with values specified in manuals to determine if the characteristics of the transistor under test are really within the limits set by its manufacturer.

Any of the testers can be used to determine if the transistor is operating as a transistor should. If there is beta or transconductance, even if you do not know what it should be, you may assume that the transistor is good. The instrument in Fig. 4-12 does this with a good–bad test, without really indicating a beta or transconductance reading. It is designed not so much to provide data, but to reveal to you rapidly if the device is good or bad. That is usually all the information that the serviceman needs, and needs quickly. More device characteristics can be determined from readings that can be made on the previously mentioned testers, but measuring procedures are somewhat more complex. So if price is not the prime consideration, determine which factor is the more important for you—speed or data—and maybe get two (or all three) testers.

Tube Tester

In this volume, the discussion centers around transceivers using semiconductor devices, although much of the information can be applied to older sets using tubes. If you should come across a CB radio with tubes, you will find a tube tester a convenient instrument. Although instruments providing only indications of cathode emission are useful, those measuring dynamic conductance, such as the B & K 747B or Eico 667, are more indicative of vacuum-tube quality.

The best method of checking a tube is, of course, by substituting a known good one. If a tube is not available as a substitute for each type in a transceiver, a tester must, of course, be used.

Oscilloscope

There are many instruments that you can learn to live without, but the cathode-ray oscilloscope does not fall into that category. Used properly, the oscilloscope can become the most valuable instrument on the service bench. It certainly is the most versatile. Anyone who has learned to use an oscilloscope properly and to utilize all the features of the instrument considers it to be indispensible when handling CB radios.

The heart of the oscilloscope is the cathode-ray tube (CRT). Electrons emitted by the heated cathode in the tube are attracted to a phosphorous screen. Formed into a thin beam by fields in the tube, the electrons appear as a visible dot upon striking the screen at its midpoint. Plates in the CRT are positioned to deflect the electron beam and dot in either a vertical or horizontal direction, or in both directions simultaneously when a voltage is applied to them. Should an alternating voltage be applied to horizontal plates, the electron beam will be deflected horizontally by different amounts during the cycle. Because of the time lag (persistence) of the fluorescent material on the screen, a continuous horizontal line will appear. Similarly, if ac voltage is applied to the vertical deflection plates, a vertical line will appear. Tilted lines, circles, or other patterns will be displayed if signals are applied simultaneously to both plates.

To get substantial amounts of deflection when minute voltages are applied to the inputs, vertical and horizontal amplifiers are built into oscilloscopes. The gain of these amplifiers is adjustable through controls on the front panel. Outputs from the amplifiers are applied to the vertical and horizontal plates in the CRT.

An oscillator generating signals in the shape of a sawtooth is built into every oscilloscope. Unlike sinusoidal signals, a sawtooth voltage increases linearly throughout the cycle. For example, if the voltage is zero at the beginning of the cycle and 2 V at the end of the cycle, it is 1 V when half the cycle has been completed. The sawtooth voltage is applied through a switching circuit to the horizontal amplifier, deflecting the electron beam on the screen in a horizontal direction. The frequency of the sawtooth can be changed by adjusting controls on the front panel of

Oscilloscope

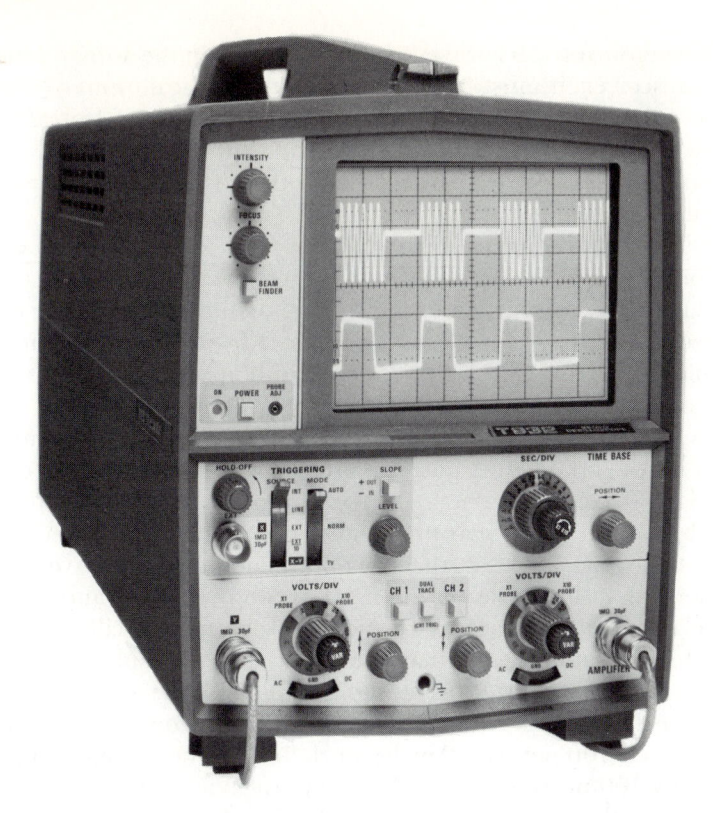

FIG. 4-15   Wideband oscilloscope. (Courtesy Tektronix, Inc.)

the oscilloscope. The horizontal deflection voltage varies linearly with time. Should alternating current be applied at the vertical input while the sawtooth is sweeping the beam in the horizontal direction, the waveshape of the ac signal is displayed on the screen.

The scope can be used to determine the frequency of a signal using a *Lissajous pattern*. Simply apply the signal at the unknown frequency to the horizontal input of the oscilloscope and the signal from a calibrated variable-frequency generator to the vertical input. Adjust the generator until either a circle, straight line, or ellipse is on the screen. The frequency produced at the generator is then identical to the frequency in question.

Percent of modulation can also be observed from patterns on an oscilloscope. To do this, your oscilloscope must be able to reproduce RF frequencies properly. The relatively inexpensive Tektronix T932 shown in Fig. 4-15 has frequency-amplifying capabilities to more than 35 MHz. It can easily be used to display signals in the 27-MHz CB band.

To monitor the RF waveshape of the signal in the antenna circuit of a CB transceiver, it must be connected to the vertical input of the oscilloscope. The load presented by the oscilloscope must not affect the shape or amplitude of the signal. To assure this, adaptor LA-31 in Fig. 4-16, sold by Leader, may be placed in the circuit between the transceiver and antenna.

Using a short coaxial lead (similar to the lead described in Chapter 3 for connecting an SWR meter to the transceiver), connect the antenna output from the transceiver to the RF jack on the adaptor; then connect the antenna (or, better yet, connect a 52-Ω load) to the remaining RF jack on the adaptor. Connect the RF vertical input lead from the oscilloscope to the terminals on the front of the adaptor, observing that the ground connector from the oscilloscope is wired to the terminal so labeled on the adaptor. Adjust the oscilloscope so that the trace on the screen is deflected horizontally by the sawtooth voltage. Press the push-to-talk button on your microphone and whistle into it. You should get a pattern as in Fig. 2-8c. Set the sawtooth generator so that you can observe several audio cycles on the display. Stop whistling and adjust the RF level (using either the vertical gain controls on the oscilloscope or the control on the adaptor) so that you get a vertical display of two large divisions on the screen. Whistle into the microphone again. If peaks in the audio deflect the RF to four large divisions on the CRT screen, modulation is 100 percent. Any lesser deflection means that modulation is less than 100 percent. Methods used to calculate the percent of modulation from the modulated waveshape were discussed in Chapter 3.

Trapezoidal patterns are frequently used to determine percent of modulation. To produce the trapezoidal pattern on the screen, feed the modulated signal to the vertical input as before. Instead of applying the sawtooth signal from inside the oscilloscope to the horizontal amplifier, connect the input of this amplifier to the audio output from the modulator inside the transceiver. The meaning of the various trapezoidal

FIG. 4-16 Scope adaptor. (Courtesy Leader Instruments Corp.)

Oscilloscope

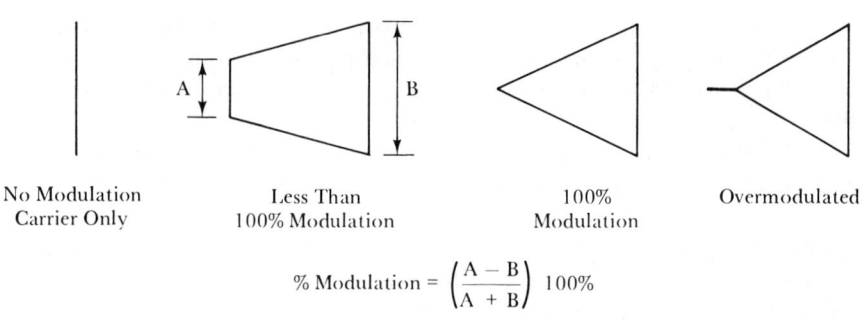

| No Modulation Carrier Only | Less Than 100% Modulation | 100% Modulation | Overmodulated |

$$\% \text{ Modulation} = \left(\frac{A - B}{A + B}\right) 100\%$$

FIG. 4-17 Trapezoidal patterns to determine modulation.

patterns and the formula for calculating the actual percent of modulation are given in Fig. 4-17.

Most oscilloscopes do not have amplifiers with sufficient bandwidth to display the 27-MHz RF of the CB band. The modulated display is still possible on many oscilloscopes if the adaptor in Fig. 4-16 is connected directly to the deflection plates of the CRT. Combined into one unit with an adaptor for RF signal display, the Leader oscilloscope in Fig. 4-18 is a convenient instrument specifically designed for checking AM as well as SSB.

Any oscilloscope can be treated as a voltmeter. Should a sinusoidal signal be applied to the vertical input while a sawtooth of the proper frequency is at the horizontal, the relative voltages at each instant in the cycle are displayed. Voltage from the crest to the peak of the sine wave is the peak-to-peak voltage.

If the oscilloscope features vertical dc amplifiers, the zero dc level is

FIG. 4-18 Scope for RF applications. (Courtesy Leader Instruments Corp.)

determined by shorting the vertical input leads to ground. A straight horizontal line is displayed. Any dc voltage applied to the leads will deflect the line either above or below the zero level, depending upon the polarity.

When used as a voltmeter, as a device to display waveshapes, or as an indicator of dc voltage levels, the oscilloscope can serve a very important function in tracing signals through the transceiver. Should the RF frequency be beyond the capability of the oscilloscope, an RF probe identical to the one shown in Fig. 4-4 can be used here. It will convert the RF to direct current. Peak ac will be displayed on the screen as a dc level. However useful this information may be, a look at the actual waveshape is more informative. But true RF waveshapes can only be displayed on the more expensive wideband oscilloscopes. Before purchasing an oscilloscope, determine if it will serve your purpose in displaying a sufficient band of frequencies for CB troubleshooting applications. You can undoubtedly live comfortably with either instrument shown here.

Signal Tracer

One of the most convenient and rapid ways of troubleshooting a receiver is by signal tracing. Feed a modulated signal to the input. Then check for signal at each stage, beginning at the RF amplifier, through the converter, IF and detector stages, and finally at the audio output via the voltage amplifiers. The presence of signal at the output circuit of each stage can be determined either from a display on an oscilloscope, or through the use of a signal tracer such as the Eico instrument shown in Fig. 4-19. The signal tracer is supplied with two passive probes, one for tracing RF signal through these stages, and the other for observing the performance of the audio section.

An inexpensive signal tracer probe, complete within itself, is also produced by Eico (see Fig. 4-20). The amplifier is built into the probe case. Three probe tips are supplied, one for demodulating an RF signal and two for checking audio stages. The two audio tips have built-in attenuators for observing amplifiers capable of producing a wide variety of signal levels. A plug earphone is used as the transducer to let you hear if signal is present at the various stages.

Both signal tracers and oscilloscopes can be used for observing the presence of modulating audio signal and the modulated RF at the various stages in the transceiver. Wideband oscilloscopes must be used to observe unmodulated RF. Signal tracers can perform their function by checking the various stages only after the RF has been modulated.

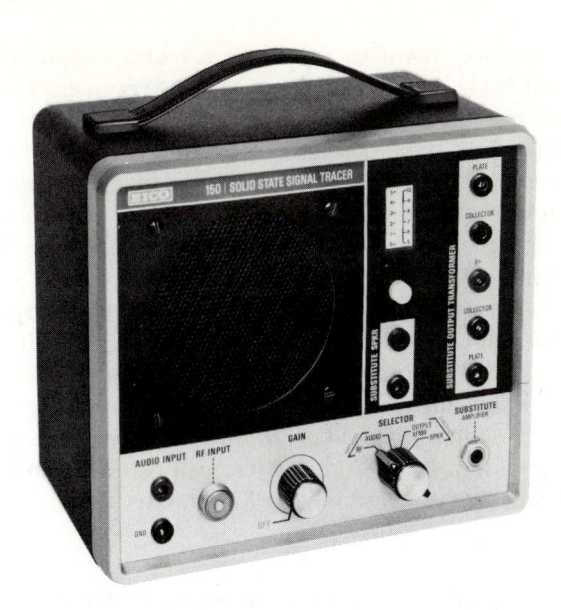

FIG. 4-19 Signal tracer. (Courtesy EICO Electronic Instrument Co., Inc.)

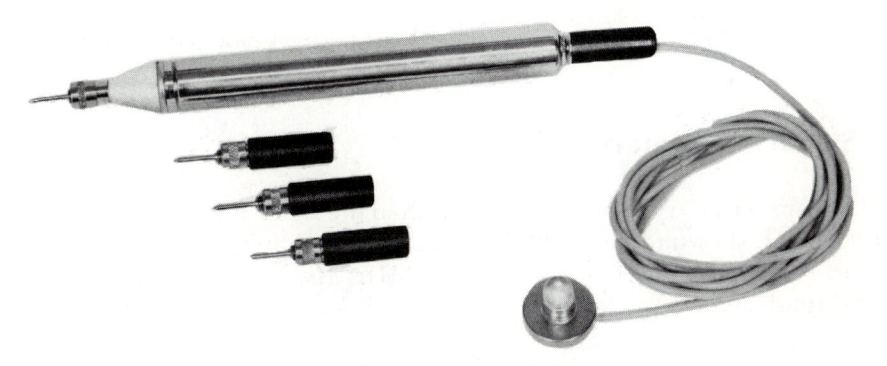

FIG. 4-20 Signal tracer probe. (Courtesy EICO Electronic Instrument Co., Inc.)

Signal Injector

Signal tracing is performed by observing the presence of signals in sequence, beginning at the input of a receiver and proceeding by checking stages in sequence to the output. However unlikely, the actual defect in the equipment may elude you when you do this. Other procedures are called for.

One method involves injecting a 1-kHz signal with strong harmonics to 30 MHz into the final stage of the receiver. This type of signal is generated by the Eico PSI-1 signal injector probe shown in Fig. 4-21. If signal injected into the last stage is reproduced through the loudspeaker in the transceiver, the stage is assumed to be operating properly. The probe is then moved to each proceding stage in sequence, one stage at a time, injecting the 1-kHz signal loaded with harmonics. The same signal is injected into every stage, regardless of whether it was designed to amplify RF, IF, or audio. If the stage is operating properly, it will amplify the fundamental signal as supplied by the probe or one of the harmonics. If all stages between where the signal is injected into the receiver and the audio output are operating properly, signal will be reproduced by the loudspeaker. Should the signal be injected into the input circuit of a defective stage, there will be no audio emanating from the loudspeaker, or the audible level will be drastically reduced from the norm.

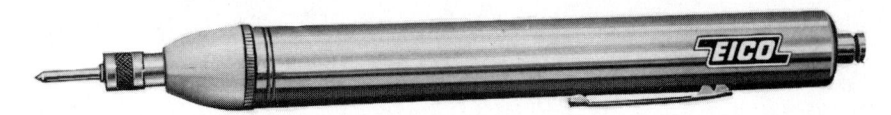

FIG. 4-21   Signal injector probe. (Courtesy EICO Electronic Instrument Co., Inc.)

Signal Generators

A more serious type of RF generator than the signal injector probe is the Eico unit shown in Fig. 4-22. It can be used to supply pure fundamental or modulated RF frequencies to 54 MHz. Available signals are more than adequate for use in all CB troubleshooting and alignment procedures.

Crystal oscillators in transceivers are frequently at fault when the receiver or transmitter section does not function. Proper performance may be restored by injecting a signal from the RF generator into the affected circuit of the transceiver (probably at the mixer stage) as a substitute for signal from the crystal oscillator. Any improvement in reception due to this procedure is positive proof that the crystal oscillator is defective. The crystal itself is usually the defective component in the circuit.

A more important function of a signal source is to supply RF for aligning the receiver and transmitter. Obviously, you cannot rely on the frequency calibration etched into the dial on the instrument. It is not sufficiently accurate by itself to meet the stringent requirements of the

Signal Generators

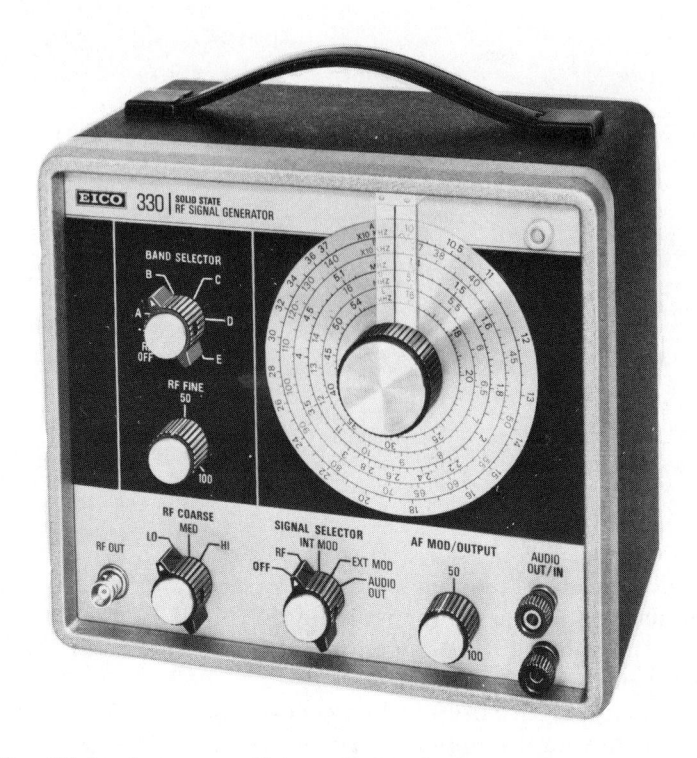

FIG. 4-22 RF signal generator. (Courtesy EICO Electronic Instrument Co., Inc.)

FIG. 4-23 Frequency counter. (Courtesy B&K Dynascan Corp.)

FCC. A frequency counter, such as shown in Fig. 4-23, should be used to monitor the RF output frequency from the generator.

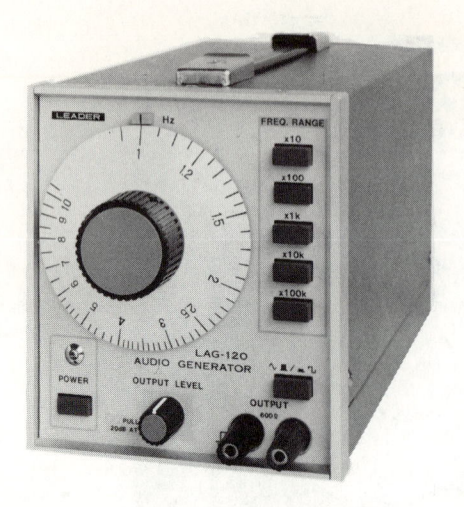

FIG. 4-24 Leader LAG-120 audio signal generator. (Courtesy Leader Instruments Corp.)

Procedures for aligning the receiver section of the CB radio are very similar to those used when aligning ordinary AM radios. A signal generator must be used to supply the modulated RF and IF signals for alignment purposes.

Audio signals are useful when checking squelch circuits, as well as for testing the audio amplifier section of the transceiver. Many RF generators will provide some type of audio signal for this purpose. Audio from the generator can be used in several instances when troubleshooting the transceiver. It may be used as a substitute source of signal to modulate the transceiver in the transmitting mode of operation, if the modulator in the transceiver is defective. It is also useful when adjusting the limiter circuit in the transceiver.

A good RF generator is essential. Considerably less important in service work, but nice to have, is an audio signal generator, as shown in Fig. 4-24, supplying sine and square waves at frequencies from 20 Hz to 20 kHz. It is useful for checking the frequency response of the audio amplifier section of the rig when used for reproduction of the received signal, as well as when modulating the transmitter section of the CB radio. The transmitted bandwidth of the transceiver can only be determined if an instrument of this type, supplying a wide range of audio frequencies, is used to feed signal to the modulator.

The public-address (PA) feature commonly found on transceivers involves only the audio circuits. Power microphones use only audio circuits. Frequency characteristics and distortion of these amplifiers can best be checked and evaluated when using a good audio generator.

Let us stress once again: although it is frequently convenient and useful to have an audio generator, and essential when you are working on hi-fi equipment, its applications to CB servicing are very limited.

Digital Frequency Counter

The best method of checking a frequency is through the use of the modern digital counter, such as the B & K instrument shown in Fig. 4-23. Although designed to cover a frequency range of from 20 Hz to 40 MHz, the typical production instrument can count accurately from 10 Hz to 60 MHz. Either limit is more than sufficient in CB applications.

The B & K 1801 instrument has a unique overflow feature, which expands what seems a limited display of but six digits into a realistic eight-digit display. When looking at the instrument, you see only six digits. This is somewhat limited and confining when working in the megahertz range, for there is a distinct possibility that eight significant digits may be required. For example, the frequency 27.169413 MHz or 27,169,413 Hz has eight digits. In a six-digit display, all you would normally see is 27.1694 MHz. The expansion feature consists of the means to suppress the first two significant digits (27 in this example), so that you see only the last six digits. Thus from one reading in our example you know the frequency is somewhat greater than 27 MHz. You can determine the last six digits from the second reading as 169413. As on all counters, the last or least significant digit can be wrong by one count, so the last digit may be either a 2, 3, or 4.

FCC standards require that the transmitted frequency be accurate within ±0.005 percent. At 27 MHz, this means that the RF signal must not vary by more than 0.00135 MHz (or 1350 Hz) above or below the assigned frequency. The B & K instrument is accurate to within 0.001 percent. At 27 MHz, this is ±270 Hz. Because the last digit can be wrong by 1 count, the instrument is accurate to within 271 Hz of the indicated reading. To be certain that you are within FCC frequency limitations when measurements are made with this instrument, you must account for the ±271-Hz inaccuracy. You cannot use 27 MHz ±1350 Hz as a reading that will be acceptable to the FCC. You must limit the 1350-Hz tolerance by the possible 271-Hz error introduced by the instrument to ±1079 Hz. Thus, if the instrument indicates 27 MHz ±1079 Hz, or any frequency between 26.998921 MHz and 27.001079 MHz, the frequency delivered by the transceiver that should be at 27 MHz meets FCC requirements.

The Hickok 388 frequency counter shown in Fig. 4-25 features a seven-digit display. This is a satisfactory number of digits in most instances, so the overflow feature is not necessary here and has not been

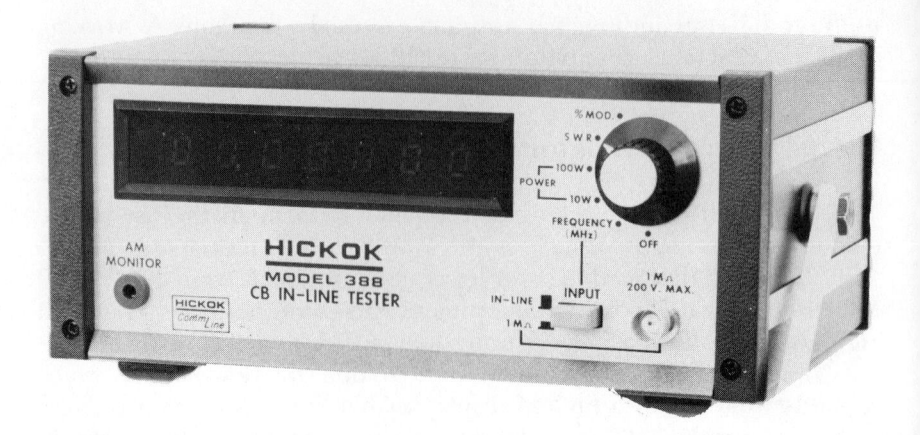

FIG. 4-25 Frequency counter and in-line RF tester. (Courtesy Hickok Electrical Instrument Co.)

designed into this instrument. The one-count error applies here to the seventh digit as it is the least significant digit.

Two versions of this instrument are available. One has a 0.001 percent accuracy, and the other more expensive version is accurate to 0.0001 percent. We shall consider the former instrument, as 0.0001 percent accuracy is much more than is actually required.

As before, 0.001 percent accuracy indicates a possible readout error of ±270 Hz. Because seven digits can be displayed, the one-count error is now in the ten's digit, so the total readout error is ±280 Hz. Considering the ±1350-Hz tolerance allowed by the FCC for the 27-MHz signal, the ±280-Hz inaccuracy of the instrument reduces the acceptable readout to ±1070 Hz around the 27-MHz center frequency. Any reading of 26.99893 through 27.00107 MHz on the seven-digit Hickok counter is acceptable as far as the FCC is concerned. The slight tightening of transmitter frequency limits due to the digital instruments is insignificant, and will not cause any problem or excess restrictions in servicing procedures.

The Hickok 388 has several conveniences not readily found on other counters, such as a built-in adaptor similar to the one in Fig. 4-16, so that the antenna line from the transceiver can be connected through the counter. Cycles of transmitted RF frequencies can thus be counted without being forced to use more complicated procedures. Model 388 counter also has facilities to let you read SWR, output power in two ranges (10 and 100 W), and percent of modulation from the digital display.

A demodulator is built into the model 388. Demodulated signal can be observed on an oscilloscope connected to output terminals on the counter. The shape of the audio signal can be observed after it has been "processed" through modulation in the transceiver.

Once again, you have a choice of two types of instruments to do the job. Either one will serve your purposes well.

Dip Meter

We have been discussing generators and counters. Precision was assumed to be the vital factor. It is extremely important in many servicing procedures. While troubleshooting, it is frequently sufficient to have only *order-of-magnitude* information, in which case the dip meter, such as the Leader LDM-815 shown in Fig. 4-26, is useful. A dip meter embodies a resonant circuit tunable over a large RF frequency range. One element of the resonant circuit, the coil or inductance, protrudes from the instrument. The usefulness of this instrument is due to the RF fields radiated and absorbed by this coil.

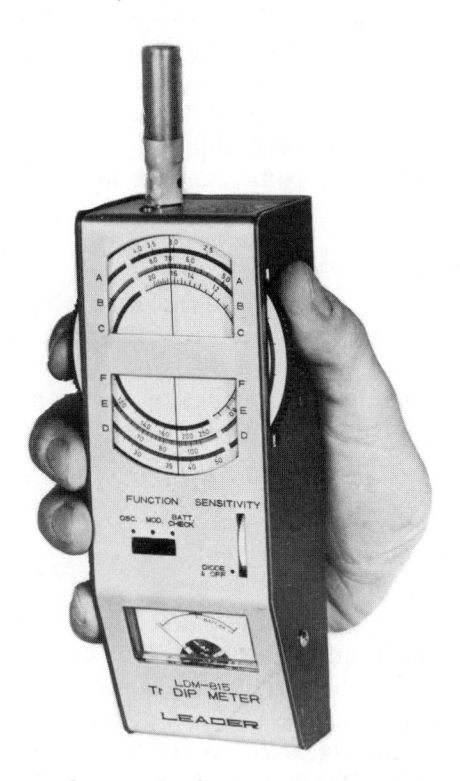

FIG. 4-26 Dip meter. (Courtesy Leader Instruments Corp.)

In one mode of operation, the tunable circuit is a component of an RF variable-frequency generator. The frequencies generated in the dip meter may be read on a dial mounted on the panel of the instrument. Because of the wide range of frequencies that are generated in the Leader instrument, it can be used to check circuits designed to operate from 1.5 to 250 MHz.

An RF voltmeter is an integral part of the instrument. When used as a signal generator, the pointer on the meter deflects. Deflection is proportional to the RF voltage available in the dip meter.

The prime function of a dip meter is to determine the resonant frequency of a circuit. If the protruding coil is placed near or loosely coupled to an unexcited resonant circuit, it will absorb some of the energy generated in the dip meter. RF voltage from the generating circuit is reduced. The pointer on the meter movement will dip. The dip is most pronounced when the frequency generated by the instrument is the same as the resonant frequency of the circuit under test. The resonant frequency of the circuit or, more precisely stated, the frequency at which the largest dip occurs can be read from the dial on the dip meter.

When the dip meter is not in an oscillating mode of operation, the protruding coil can be used to absorb energy from a circuit that is emitting RF energy. RF signal present in a circuit under test induces some of its energy into the coil on the dip meter. If the amplitude of the signal is sufficient, enough RF may be induced into the dip-meter circuit by the protruding coil to deflect the pointer on the meter. Deflection is at a maximum when the dip meter is tuned to the frequency of the energy coupled to it. As before, the frequency can be read from the dial on the panel of the instrument.

Instruments of this type are useful for locating a defective resonant circuit or an inoperative stage in a transceiver. They can also be used to check the resonant frequency of the CB antenna. For its price, no service shop should deny itself the convenience of having a dip meter at its disposal. Used properly, the instrument will pay for itself handsomely in time saved.

Radio-Frequency Wattmeter

The basic function of the transmitter section of the CB radio is to generate RF power for radiation into the air. Many instruments are available to measure RF power, but the Bird Model 43 shown in Fig. 4-27 has for many years been accepted as the standard of the industry. When used with different and interchangeable plug-in elements at the front panel, this instrument is capable of indicating RF power to 5000 W over a 2- to

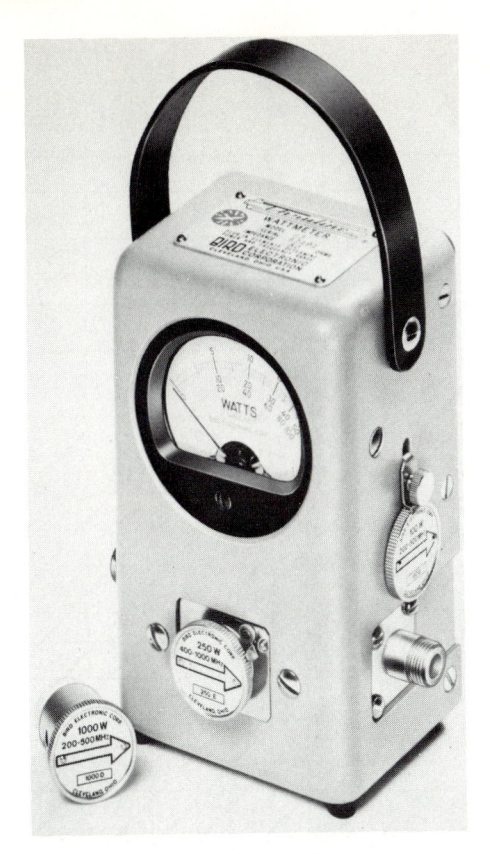

FIG. 4-27 RF power meter. (Courtesy Bird Electronic Corp.)

1000-MHz frequency range. For our purposes, plug-in element 10A is ideal. When used, and with the instrument connected in the antenna line, the meter will indicate up to 10-W power with a 5 percent accuracy factor on a frequency range of 25 to 60 MHz. The FCC limit of 4-W maximum root-mean-square (rms) output power from the transceiver can readily be measured on this instrument.

Power delivered by any type of signal, whether RF or audio, must be measured across a load. Most transceivers are designed to work into a 50- or 52-Ω load. To measure output power under ideal conditions, connect the transceiver to the jack on the left side of the wattmeter and the 52-Ω load to the jack at the right. The Gold Line resistor load shown in Fig. 3-2 mounted in a PL-259 connector is applicable here. An antenna can also be used as a load, but unless its impedance at 27 MHz is about 52 Ω, the power delivered to it will differ from that delivered to the resistor.

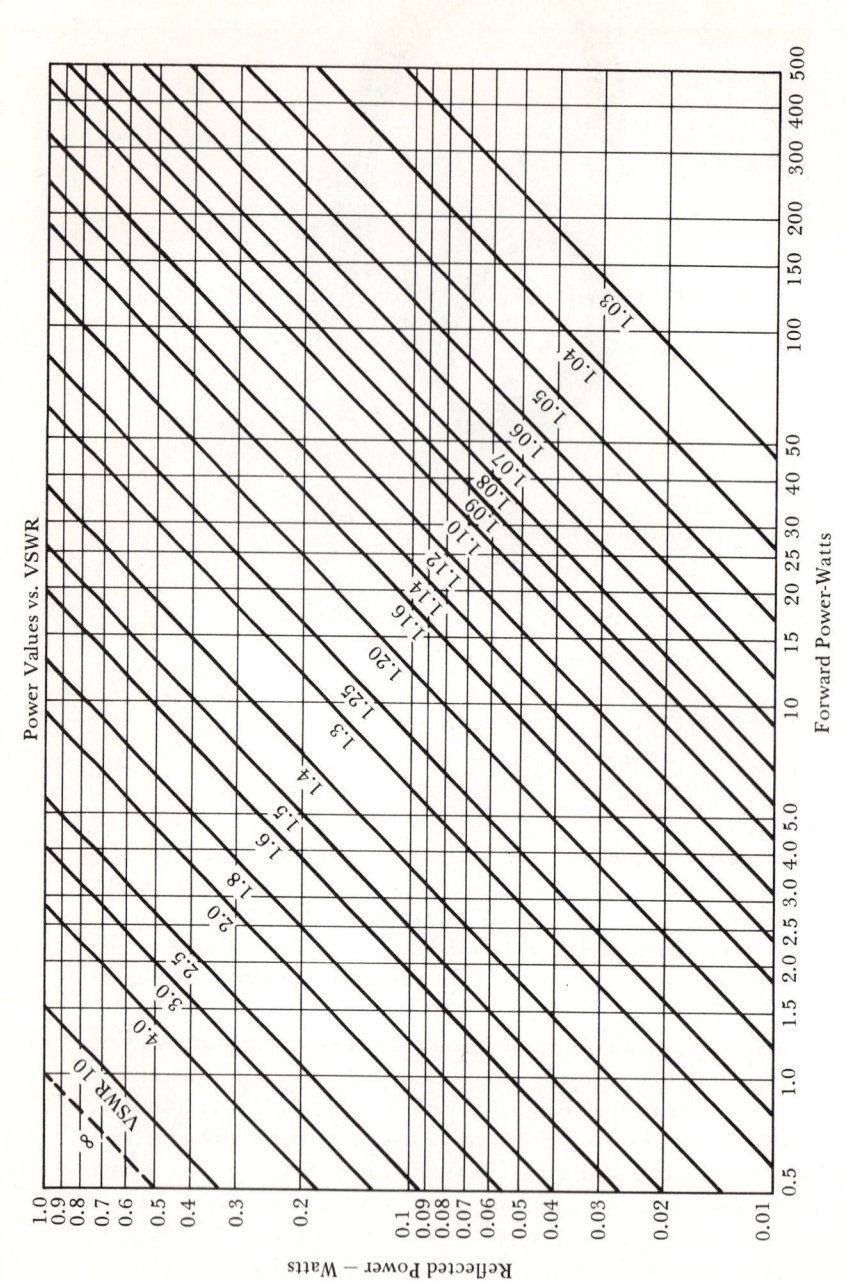

FIG. 4-28 Determine SWR from forward and reflected power measurements using these curves. Draw a vertical line from the Forward Power axis and a horizontal line from the Reflected Power axis. They cross on or near a line on the chart indicating a locus of points for a particular SWR. The SWR of the system is the number indicated by the particular line on the chart. (Courtesy Bird Electronic Corp.)

Special Instruments for Use in the Field 85

Note the arrow on the plug-in element. When oriented as shown in the figure, the meter pointer will indicate the forward power delivered by the transceiver to the antenna or 52-Ω load. The arrow can also be oriented so that it is pointing in the opposite direction. Readings on the meter will then indicate power reflected from the antenna (or resistor load) back to the transceiver due to a mismatch between the transmission line and antenna impedances. The ratio of the forward to the reflected power is related to SWR. Curves indicating this relationship are supplied with each Bird instrument and are shown in Fig. 4-28. The model 43 can thus be used as a standard when determining the SWR of an antenna system, as well as when measuring the power delivered by the CB radio.

Average power can be measured only when there is an RF carrier. There is no RF carrier in the SSB system of broadcasting. RF signal is present in pulses, and these pulses are there only when the RF is being modulated by the audio. The amplitude of these pulses is dependent upon the strength of the audio signal originally used to modulate the RF. If there is no audio, there are no RF pulses. Large pulses are present if the audio signal is strong. Because the pulses vary in duration time and amplitude with the modulating signal, all that can be measured is the peak pulse amplitude.

Legal limits have been set at 12 W peak for SSB signals by the FCC. Although the Bird Model 43 is designed to measure the average power of an RF signal, it can also be used for measuring the peak power due to SSB transmission. First, replace the plug-in element with one that will extend the meter range to more than 12 W on the 27-MHz RF band. Next whistle in the microphone until the sinusoidal shape of the RF distorts (flat tops) or until the meter reading cannot be exceeded by whistling any louder. Do not do this for more than a few seconds or you may damage the RF power output stage in the transceiver. The pointer on the meter indicates the peak power that the SSB transceiver can deliver.

The Bird Model 4314 looks very much like the instrument shown in Fig. 4-27. Continuous-wave power can be measured and read from the meter indication as on the model 43; peak power measurements can be made from pulses present in ordinary speech. Because it has a built-in power supply, the 4314 is ideal for use on the test bench.

Special Instruments for Use in the Field

Most instruments described here are for use on the test bench as well as in the field. Instruments offering the ultimate in precision when measuring various factors are usually too delicate or clumsy to be used in the field. Precision is not expected from portable instruments. Yet they must be good enough and versatile enough to provide the facility to do the installation and on-the-spot servicing job efficiently and properly.

The multimeter in Fig. 4-3 has all the measuring ranges required for checking battery and automobile regulator voltages, as well as the resistance of the antenna system to ground or to the chassis of the vehicle on which it is installed.

An SWR meter, field-strength meter, and modulation meter must be in your caddy. The Mura CBM-40 in Fig. 3-9 is excellent for checking the antenna system and the percent of modulation. The latter measurement will be discussed further in the chapter on microphones where adjustment procedures for power microphones will be described. SWR and modulation were discussed above.

Throw a Gold Line 52-Ω dummy load into your caddy, and it is now complete with all the measuring equipment to tackle any job in the field. By the way, don't forget your hand tools!

5

Troubleshooting the Transmitter

No limits have been set by the FCC as to who is permitted to troubleshoot a receiver. This leniency is acceptable, because even with numerous defects an improperly serviced receiver is not capable of outrageous interference with the various radio services. Transmitters, on the other hand, when handled improperly, can wreak havoc over portions of the RF spectrum. Consequently, anyone servicing the transmitter section of a CB radio must have at least a second-class radiotelephone license from the FCC.

Transmitters and receivers were discussed nontechnically in Chapter 2. A block diagram of the transmitter section by itself is shown in Fig. 5-1. All transmit–receive switching circuits have been deleted from the drawing as we are primarily concerned here with one mode of operation—the transmit mode.

RF is generated by a crystal-controlled oscillator, amplified by the driver and RF power amplifier, and fed to the antenna for radiation into

87

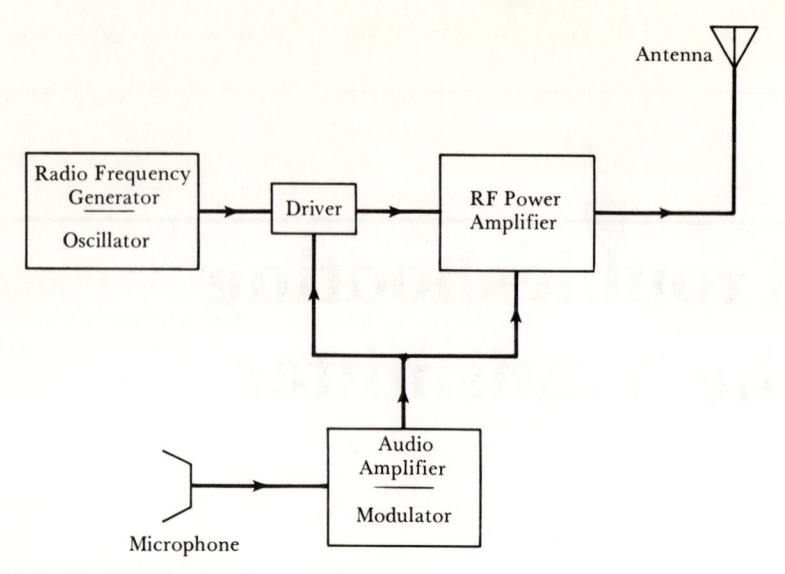

FIG. 5-1 Block diagram of transmitter section of CB radio using standard AM (amplitude modulation) techniques.

the atmosphere. The driver (or buffer stage) amplifies the RF while isolating the oscillator from the RF power amplifier to assure good frequency stability. Signals from the microphone are magnified by the audio amplifier section. Sufficient audio signal is developed to amplitude modulate the RF in the driver and RF amplifier stages.

Amplitude-Modulation Transmitter Blocks

Different circuits are used by the various manufacturers to fulfill each function in the transmitter. Some of the more common ones will be described. Just how the circuit performs its function in the CB radio that you are servicing can usually be inferred from the general discussion. Troubleshooting techniques applied to various circuits are also described, along with details as to just what generally goes wrong.

Oscillator

It is always easy to locate the various oscillators drawn in the schematic of a transceiver. Find the crystals. They are usually shown connected as a group to terminals on a switch. Follow the leads from the switch to the nearest transistor, and you have located the basic components that compose one of the oscillator circuits in the transceiver—the crystals and the associated transistor.

Amplitude-Modulation Transmitter Blocks

Many CB radios have three oscillator circuits; two are active in the receive mode of operation and two are used while transmitting. Another circuit is added when the single sideband (SSB) type of signal is being generated in the transceiver. Designers using the new phase-locked loop (PLL) circuits applying digital techniques have as their ultimate goal the use of one oscillator circuit with only one crystal to fulfill all functions of the various oscillators in the CB radio.

Two Crystals per Channel. The earliest transceivers were designed with two oscillator circuits: (1) the local oscillator in the single conversion receiver section of the CB radio, and (2) an oscillator to generate the frequency for transmission. Each oscillator circuit has facilities for a bank of 23 crystals. One crystal in each circuit is used when the selector switch is set to a particular channel. Two crystals are thus used at each setting of the switch, one for the transmitter oscillator circuit and one for the receiver oscillator circuit. Forty-six crystals must be used if all the 23 channels permitted through 1976 are to be fully active. Because crystals are expensive, few people who own this type of transceiver use more than six channels. All but the very rich, or the really avid CBer, satisfies himself with as few as ten or twelve crystals covering reception and transmission on only a few of the CB channels.

Frequency Synthesizers. The more modern 23-channel transceivers use only 14 crystals in three oscillator circuits. Yet complete coverage of all 23 channels is provided. In the excellent single conversion Johnson Model 123A transceiver, frequencies in the 32.8-MHz range are generated by an oscillator using a group of six crystals. These frequencies are beat against those in the 6.2-MHz range generated by a second oscillator using a bank of four crystals. Sufficient combinations of these two oscillators are made available to synthesize all 23 frequencies required by the transceiver when it is in the receive mode of operation. The same six 32.8-MHz frequencies beat against four crystal-controlled frequencies in the 5.7-MHz range to produce the 23 frequencies for transmission purposes. Frequencies generated here are representative of those used in transceivers embodying single conversion receiver circuits, and they are by no means to be considered as universal standards.

Just how these two groups of frequencies are combined to establish the 23 frequencies needed to fulfill the local oscillator function in the receiver section of the transceiver, while supplying the 23 frequencies for transmission, is shown in Fig. 5-2. The mixer stage in the synthesizer is very much like the mixer in an ordinary AM radio. Two frequencies are combined in this stage to provide an amplified version of the original two frequencies, along with sum and difference frequencies. In most AM radios, only the difference frequency, usually 455 kHz, is selected by

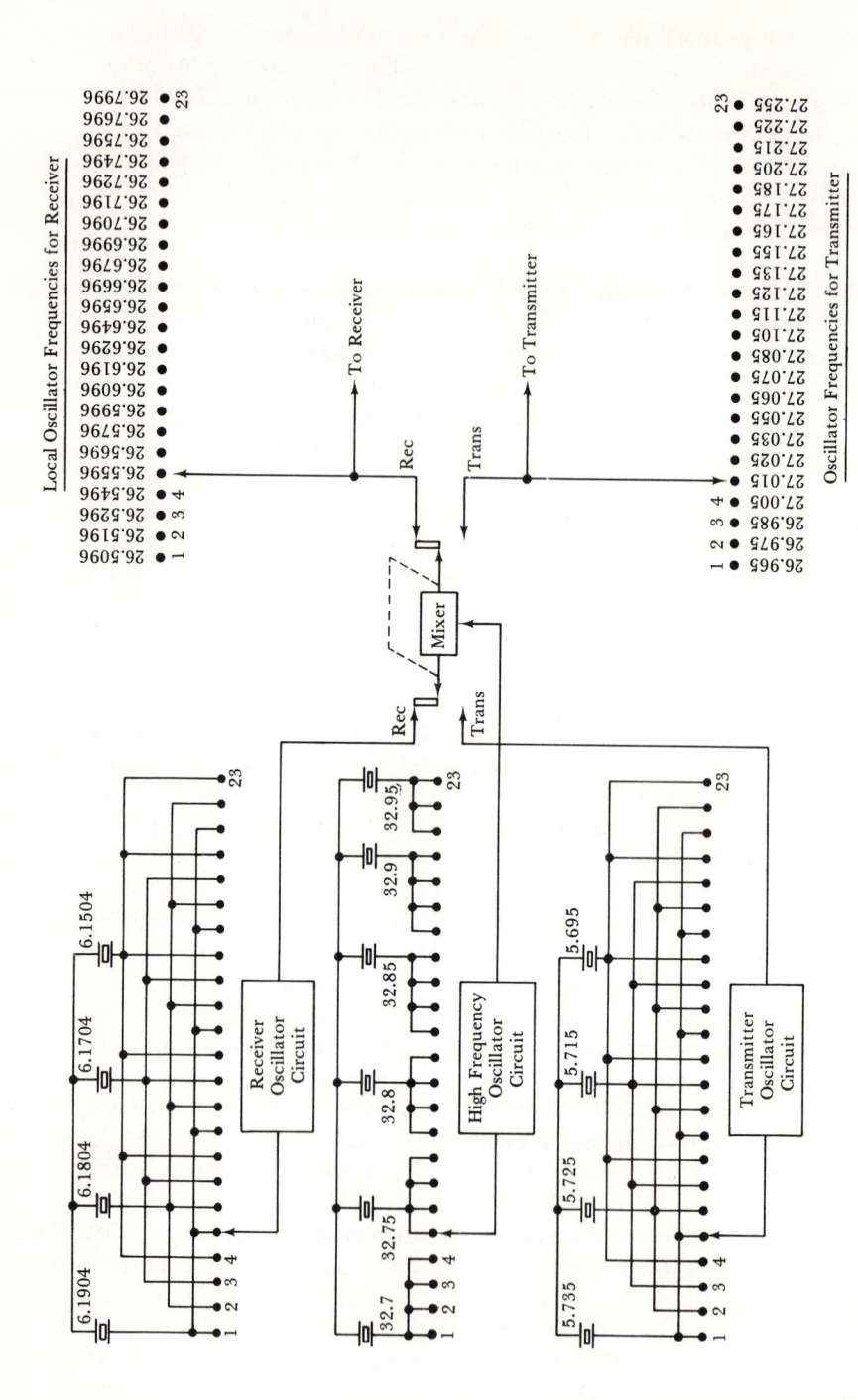

FIG. 5-2 The frequency synthesizer for 23-channel transceivers with single conversion receiver circuits. All frequencies shown are in MHz.

Amplitude-Modulation Transmitter Blocks

IF filter circuits, while all other frequencies from the mixer are discarded. In the synthesizer of Fig. 5-2, the difference frequency is once again the only one used; all others are rejected.

A 23-position, 5-pole switch is shown in Fig. 5-2. Each position represents a channel. On each pole, channels 1, 2, 3, 4, and 23 are numbered. All channels between 4 and 23 can be considered as being in sequence between channels 4 and 23 at the other positions of the switch, although the actual channel numbers are not shown.

Different functions are assigned to each of the five poles or sections of the switch. The pole at the center near the left edge of the drawing is used for selecting one of the six crystals for the high-frequency oscillator. The output from this oscillator is fed to the mixer stage. The receiver oscillator circuit using four crystals is drawn just above the HF oscillator. Similarly, four crystals are used in the transmitter oscillator shown below the HF oscillator.

Signals from the receiver and transmitter oscillators are fed through a receiver–transmitter (REC–TRANS) switch to the mixer. Here the proper signal is selected for the desired mode of operation. After the chosen signal is beat against the high frequency in the mixer, its output is directed to the receiver or transmitter circuit, as required, through a second section of the REC–TRANS switch. In the receiver, the signal is combined or mixed with the received RF to form the 455-kHz IF. Here the output from the synthesizer functions as the local oscillator. In the transmitter circuit, the signal is amplified, modulated, and radiated into the atmosphere. Transmitter and receiver oscillator frequencies must always be 455 kHz apart for the transmitted and received RF on any one channel to be at identical frequencies.

In the HF oscillator, one frequency, 32.7 MHz, is generated when the selector switch is set to channels 1, 2, 3, or 4. In the receive mode of operation, it combines with 6.1904 MHz to form the 26.5906 MHz (32.7 − 6.1904 = 26.5096 MHz) needed as the local oscillator for channel 1; it combines with the 6.1804 MHz to form the 26.5196 MHz (32.7 − 6.1804 = 26.5196 MHz) needed as the local oscillator for channel 2. The high frequency combines with the 6.1704 and 6.1504 MHz from the receiver oscillator to form the 26.5296 and 26.5496 MHz, respectively, for channels 3 and 4. Similarly, in the next four positions of the selector switch, the four frequencies generated by the receiver oscillator combine with the 32.75-MHz HF signal to form the local oscillator frequency for channels 5 to 8. This continues all the way down the line through channel 23, as shown in Fig. 5-2.

Frequencies for the transmitter are generated in an identical manner, but this time the HF signal combines with frequencies generated by the transmitter oscillator circuit.

Frequencies formed by the synthesizer are indicated at the two poles of the switch shown at the right side of Fig. 5-2. The 23 frequencies required at each switch position for the local oscillator in the receiver section are indicated at the pole of the switch drawn near the top border. At the pole drawn below this, the 23 oscillator frequencies used in the transmitter section of the CB radio are denoted at the various switch positions.

In a practical circuit, the two switch poles shown at the right of Fig. 5-2 do not exist. Proper frequencies on each channel are fed *directly* from the synthesizer to the mixer in the receiver section of the CB radio and to the driver in the transmitter section. No additional switching is needed after the mixer in the synthesizer and the three poles of the switch at the left edge of Fig. 5-2 have done their jobs.

In Fig. 5-2, one bank of four crystals is shown at the receiver oscillator circuit and a second bank of four crystals at the transmitter oscillator circuit. These two banks of different crystals must be used in the synthesizer. However, only one physical electronic circuit is needed. Here, as before, two of the crystals are selected in each setting of the 23-pole switch. Depending upon whether you are receiving or transmitting, the proper crystal is applied to the electronic oscillator circuit through a section of the REC–TRANS switch. Output from this electronic circuit is mixed, as in Fig. 5-2, with the output from the HF oscillator in the synthesizer. The difference frequency is directed through an additional pole of the REC–TRANS switch to the proper section of the transceiver.

A practical circuit using this arrangement is shown in Fig. 5-3. Six HF crystals are on switch wafer S2A. These crystals form an oscillator circuit in conjunction with transistor Q13. Transmitter crystals Y1 through Y4 and receiver crystals Y5 through Y8 are on wafer S2B of the channel-selector switch. Working with oscillator transistor Q5, these crystals are important in setting and maintaining the lower frequencies in the transmit and receive modes of operation.

In this particular transceiver, the low-frequency crystals are cut for frequencies at which the circuit is to oscillate, the fundamental oscillator frequencies. On the other hand, the HF crystals are cut for frequencies equal to about one-third that required at the output of the oscillator circuit. As the signal generated by the HF oscillator circuit is not purely sinusoidal, overtones are formed. Overtones are frequencies somewhat above those that are two, three, and more times the fundamental frequency. In the circuit in Fig. 5-3, the output tank, T7, is tuned to the third overtone. This is the frequency passed on from the HF oscillator to mixer Q14, not the fundamental. (Note that the term "harmonic" was not applied here to the crystal circuit. "Harmonic" implies a frequency that is an exact multiple of the fundamental. Crystals supply overtone frequencies that are close to multiplies of the fundamental frequency but

Amplitude-Modulation Transmitter Blocks

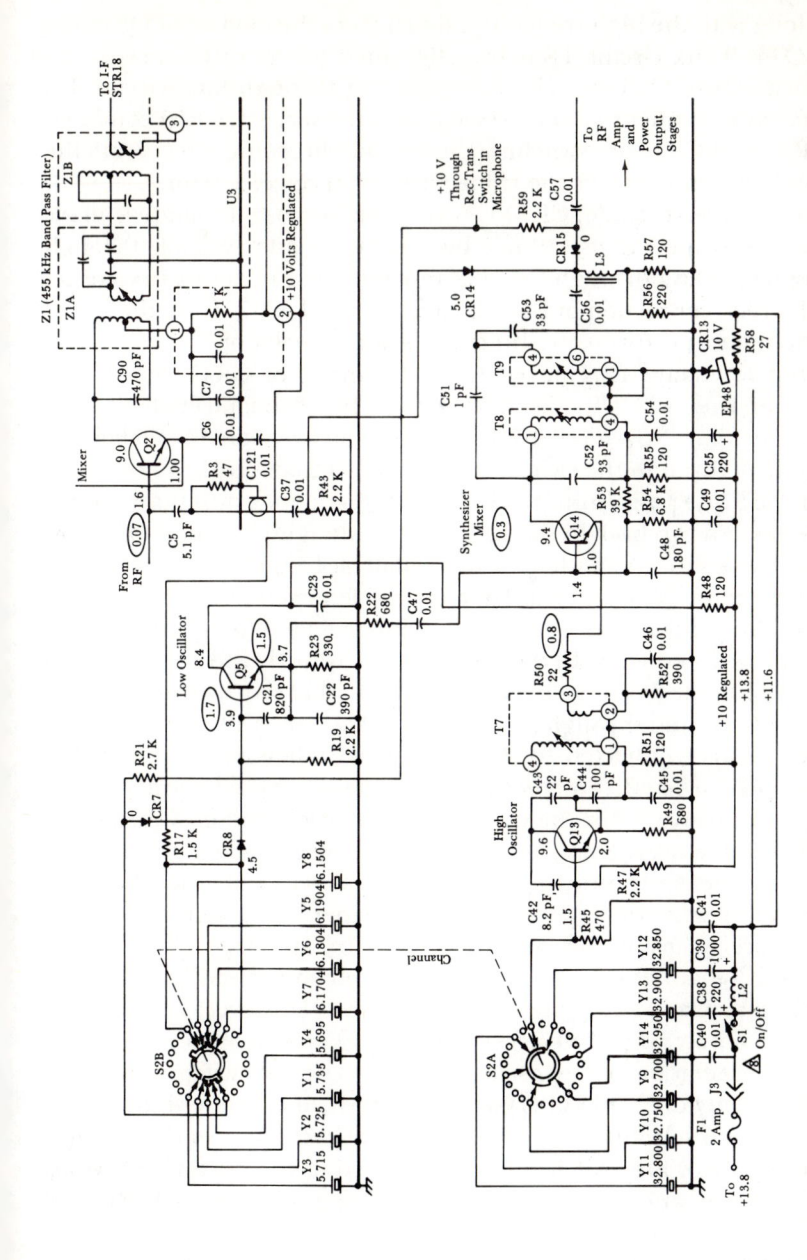

FIG. 5-3 23-channel switched frequency synthesizer using diode switching techniques. +10 volts is available only to the receiver section and diodes when the switch in the microphone is in the REC position. When switch is pushed into TRANS position, +10 volts is available only to the transmitter section and diodes. (Courtesy E.F. Johnson Co.)

are not exactly equal to these multiples. Overtone frequencies are somewhat higher than harmonic frequencies.)

Along with the high frequency, signal from the emitter of Q5 is also fed to Q14. Tank circuit T8 is broadly tuned to the difference of the frequencies from Q13 and Q5. After passing through tank circuit T9, the difference frequency is impressed across the series L3–R57 combination. CR15 and CR14 are switching diodes that direct the signal from the L–R combination to either the transmitter or receiver circuit.

In the receive mode, CR14 is turned on because its anode is made positive with respect to its cathode by a voltage made available through R43. Signal passing through CR14 is applied to the mixer or first detector in the receiver section of the CB radio through C37. Here the signal from the mixer performs the function of local oscillator.

Positive voltage is removed from the anode of CR14 and the receiver after the switch in the microphone has been pushed into the transmit position. Now diode CR15 is turned on by voltage available to it through R59 through the switch contacts. A complete path is now available for the RF to pass through CR15 and subsequent transmitter stages.

Let us now go back to the LF oscillator designed around Q5. How does this transistor know whether to oscillate at frequencies determined by the transmitter crystal or at frequencies determined by the receiver crystals? The proper bank of crystals is connected to Q5 through switching diodes CR7 and CR8. In the receive mode, diode CR8 is switched on by a voltage made available through the REC–TRANS switch circuit in the microphone and through resistor R17. Receiver oscillator crystals Y5 through Y8 are connected to the Q5 circuit. When the switch in the microphone is set so that diode CR7 is switched on by voltage at resistor R21, the transceiver is in the transmit mode of operation. Crystals Y1 through Y4 take over. Here, as in section S2A of the switch, the applicable crystal on each channel is selected by contact established between specific stator lugs through the rotating wiper. The Johnson Model 123A transceiver using this circuit is shown in Fig. 5-4.

The 23-channel Royce 1-602A transceiver shown in Fig. 5-5 uses a synthesizer similar to the one described above. However, here the receiver incorporates dual conversion circuitry. The schematic also differs in the representation of the oscillator circuit. Actual switch wafers are not shown. Only the banks of crystals are drawn, and a wiper is indicated as being able to select the applicable crystal as required. In the physical transceiver, there are switch wafers for selecting the necessary crystals for the receiver and transmitter modes of operation at each channel setting.

Crystals working with transistor Q7 are used when generating high frequencies. In this transceiver, they are at about 37.7 MHz. Receiver crystals working in conjunction with Q8 generate frequencies in the

Amplitude-Modulation Transmitter Blocks

FIG. 5-4 Johnson messenger 123A transceiver. (Courtesy E.F. Johnson Co.)

10.17-MHz range; transmitter crystals working in conjunction with Q20 provide frequencies in the 10.45-MHz range. Using channel 12 as the basis of the following discussion, let us see just what the Royce transceiver can accomplish with these frequencies, in addition to incorporating the circuits in the attractive package shown in Fig. 5-6.

When set to channel 12, the HF oscillator supplies a signal at 37.7 MHz; the frequency from the receive oscillator is 10.14 MHz, and from the transmit oscillator it is 10.595 MHz.

In the receive mode of operation, the first mixer (see Fig. 2-7) combines the 37.7 MHz from the HF oscillator with the 27.105-MHz channel 12 carrier frequency to establish the high IF of 37.7 − 27.105 MHz, or 10.595 MHz. This IF passes through the LF mixer along with the 10.14 MHz from the receive oscillator. The low IF, 10.595 − 10.14 MHz, or 0.455 MHz (455 kHz), is selected and amplified by the low IF section of the transceiver.

To fulfill the transmitting function, the 37.7-MHz high frequency from Q7 is mixed in Q19 with the 10.595 MHz from Q20. The frequency selected by the tank circuits in the collector of Q19 is 37.7 − 10.595 MHz, or 27.105 MHz. This is the carrier frequency assigned to channel 12. It is amplified by the remainder of the transmitter circuit for radiation into the atmosphere.

In the Johnson transceiver, output from the oscillator is channeled through switching diodes to the various circuits. As for the Royce 1-602A, voltage is applied to the proper oscillator and transceiver circuits as required. Consequently, the receiver section and receiver oscillator operate only when the switch in the microphone is set to the REC position. The switch must then be pushed into the transmit mode position if voltage is to be removed from the receiver electronics and receiver oscillator, and applied instead to the transmitter and its oscillator. Only the HF oscillator keeps "humming" along at all times.

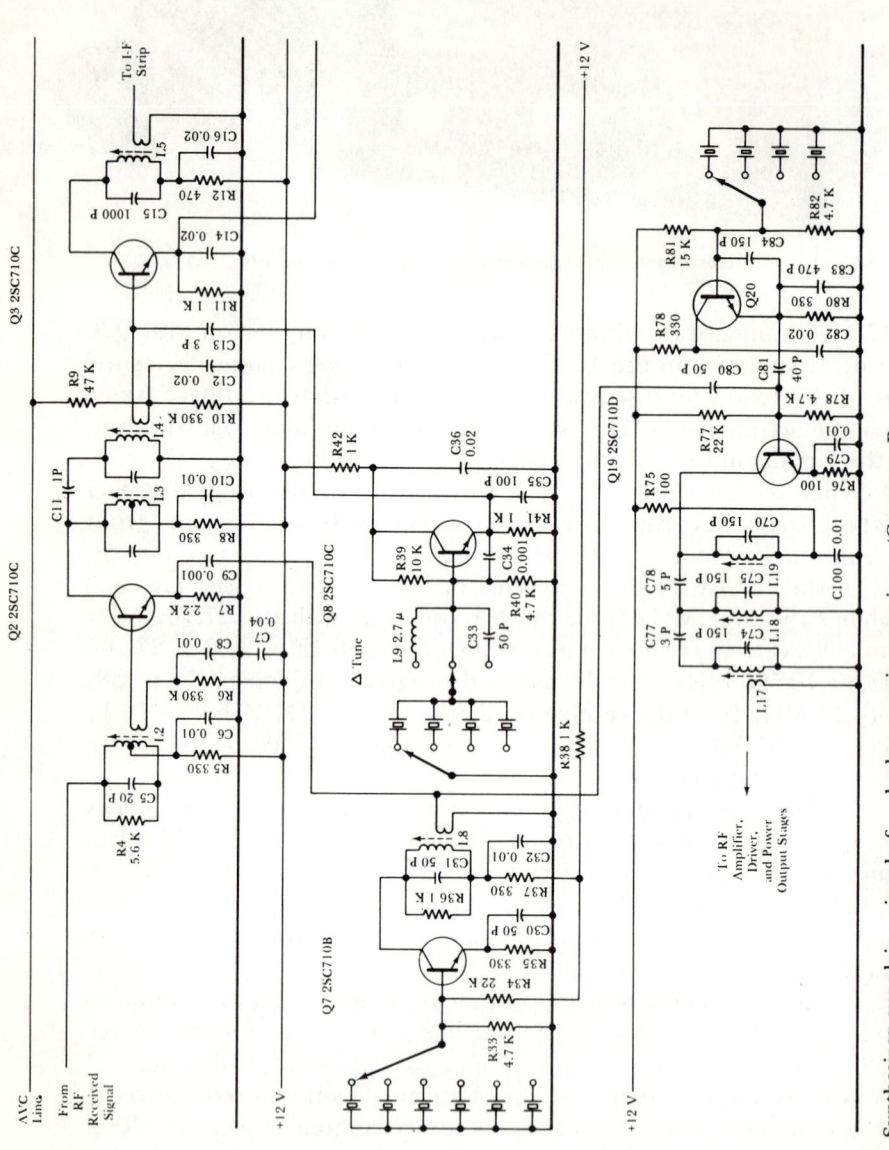

FIG. 5-5 Synthesizer supplying signals for dual-conversion receiver. (Courtesy Royce Electronics Corp.)

Amplitude-Modulation Transmitter Blocks

FIG. 5-6 Royce model 1-602 transceiver. (Courtesy Royce Electronics.)

Phase-Locked Loop. Although 14-crystal frequency synthesizers are used in most 23-channel sets on the market, the phase-locked loop (PLL) with only one crystal in a digital-type circuit is becoming the standard. Using just one crystal, many different frequencies can be generated by simply changing a voltage in the circuit. PLL is becoming popular because of the shortage of crystals, the economy of using less crystals than in the synthesizers described above, and the ease of expanding coverage to even more than the 40 channels in current use, should the FCC decide to allow this.

Numerous versions of the PLL circuit are available. We shall briefly describe the basic operation. Troubleshooting is obviously quite simple, as the entire circuit is usually supplied in integrated-circuit (IC) form. The IC can become defective; but before replacing an expensive IC, you should be certain that components around the IC are good and that all voltages applied to the IC are as specified. Only after assuring yourself of these other factors should you even venture the thought that the IC itself has a flaw.

The basic arrangement of the PLL is shown in Fig. 5-7. A stable and fixed crystal-controlled oscillator supplies a frequency anywhere between 1 and 11 MHz. Whatever the frequency supplied, it is usually divided down in frequency divider 1 to 10 kHz for CB applications. The significance of 10 kHz is that this is the spacing between CB channels.

A second important element of the PLL is the voltage-controlled oscillator (VCO). As it is to be used in CB radios, the frequencies generated here must ideally be multiples of 10 kHz. A resonant inductor–capacitor (L–C) tank is an important element of the VCO circuit. The frequency supplied by the VCO depends upon the sizes of the

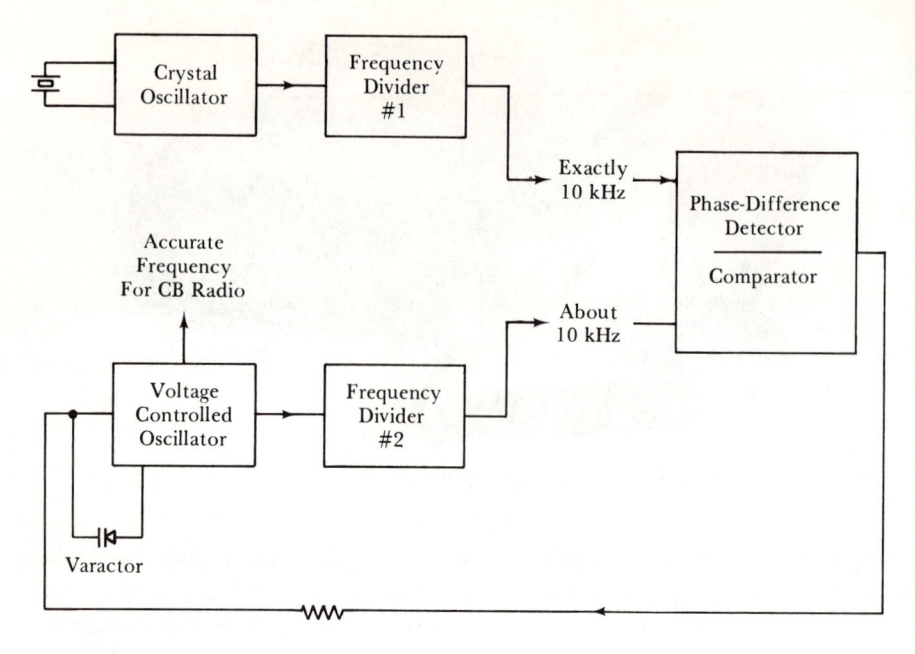

FIG. 5-7 Basic phase locked loop.

L and C in the tank. A varactor diode acts as a capacitor in the resonant L–C circuit. As the capacity of the varactor depends upon the voltage applied to it, and as the frequency of oscillation depends upon the size of the capacitor in the circuit, the frequency of oscillation depends upon the voltage applied to the varactor diode. The frequency from the VCO passes through a frequency divider. Its output is essentially at 10 kHz.

The immovable 10 kHz from the crystal oscillator is compared with the approximate 10 kHz from the VCO in the phase-difference detector circuit. Any frequency difference produces an error voltage at the output of the comparator. This voltage is fed back to the varactor of the VCO, correcting its frequency. Any time the frequency of the VCO strays off course, it is immediately corrected. The frequency of the output from the VCO is extremely stable and suitable for use in CB radios.

Frequencies delivered by the VCO can be altered by changing the constants in its L–C circuit. These frequencies are chosen so that they are useful in every section of the CB radio to satisfy all transmitting and receiving requirements. As the frequency from the VCO is altered, the percentage of the frequency from the VCO that is to be supplied by frequency divider 2 must be changed. There must be 10 kHz at its output, regardless of the frequency that is supplied by the VCO.

One practical arrangement using the PLL in a CB radio is shown in Fig. 5-8. Three crystals are used in this circuit. One crystal is in the PLL

Amplitude-Modulation Transmitter Blocks

oscillator, a second is in a fixed transmitter oscillator supplying 10.695 MHz, and the third is in a fixed local oscillator generating 10.240 MHz. The latter oscillator is connected to the second mixer in the receiver section of the transceiver.

Output frequencies supplied by the PLL circuit in this 23-channel radio are anywhere between 16.27 and 16.56 MHz. This range can be expanded for 40-channel radios. The specific frequency is determined by the CB channel selected. Let us trace the operation of a circuit using the PLL to supply the frequencies necessary for one of the channels. Once again, we shall use channel 12.

The assigned carrier frequency of channel 12 is 27.105 MHz. When the channel selector switch on the CB radio is set to channel 12, the circuitry in the PLL oscillator is adjusted so that its output frequency is 16.41 MHz, or 10.695 MHz less than the carrier. The 16.41 MHz combines with the carrier in the first mixer of the receiver circuit to produce the 10.695-MHz high IF. Subsequently, the 10.695 MHz beats against the fixed 10.240 MHz from the receiver oscillator to produce the 455-kHz difference frequency for the low IF stages.

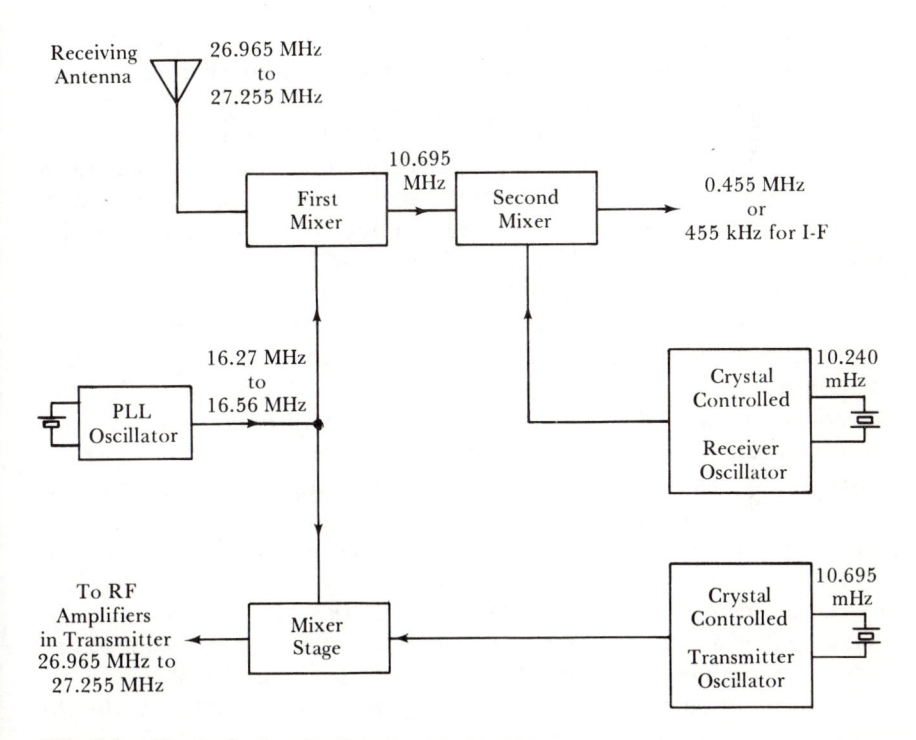

FIG. 5-8 Simple dual-conversion circuit using PLL.

In the transmitter circuit, the 16.41 MHz from the PLL is mixed with the 10.695 MHz from the transmitter oscillator to produce the 27.105-MHz RF for transmission on channel 12. Regardless of the channel considered, the output frequency from the PLL oscillator is adjusted so that the output frequency from the first mixer in the receiver section is always 10.695 MHz, and the frequency transmitted is always the same as the one being received.

A variation of the circuit just described involves a 10.240-MHz crystal oscillator as an integral part of the PLL. Frequency divider 1 in the PLL reduces this frequency before it is fed to the comparator. As before, frequencies at the output of the PLL circuit (or from the VCO) used in 23-channel radios range from 16.27 to 16.56 MHz, and are applied to the first mixer in the receiver section. The fixed frequency for the second mixer is taken from the 10.240-MHz crystal oscillator in the PLL.

In the transmit mode of operation, the frequencies provided by the PLL differ from those supplied for receiving. For 23-channel transceivers, they range from 16.725 to 17.015 MHz. These are combined in the transmitter's mixer stage with the 10.240-MHz frequency supplied by the crystal oscillator in the PLL. Proper receiver and transmitter frequencies are supplied by the PLL in the two modes of operation so that the same channel is received and transmitted when the channel selector switch on the front panel is set to a specific position.

The 23-channel SBE Model 32CB transceiver is an advanced example of a CB radio using a PLL. SBE has gone several significant steps further. They add digital logic circuitry so that channels are indicated by a digital readout. In addition, the readouts are an integral part of the microphone case along with channel selection, level, and squelch controls. Note all this updating in Fig. 5-9.

Discussion of how digital logic circuits work is not required here, as all circuitry is built around a number of ICs. If the readouts or circuit is inoperative, check the voltages at the ICs and compare them with those specified by the manufacturer. If they are not proper, first seek the defect outside the IC chip. Should it become necessary, you can replace a suspected IC. In cases where replacement of a chip returns operation to normal and all voltages at the IC terminals are as specified, you can be quite certain that the component you replaced is defective. Should the IC be soldered into the printed circuit board rather than mounted in a socket, its removal is difficult indeed. In this situation, it is best to use a solder puller to remove solder from one pin at a time. After all solder has been cleaned from the terminals and the nearby area on the printed circuit board, carefully lift the IC from the board. Be gentle. Do not break any connecting terminals in the process or you may irreparably damage what might be a good component.

Amplitude-Modulation Transmitter Blocks

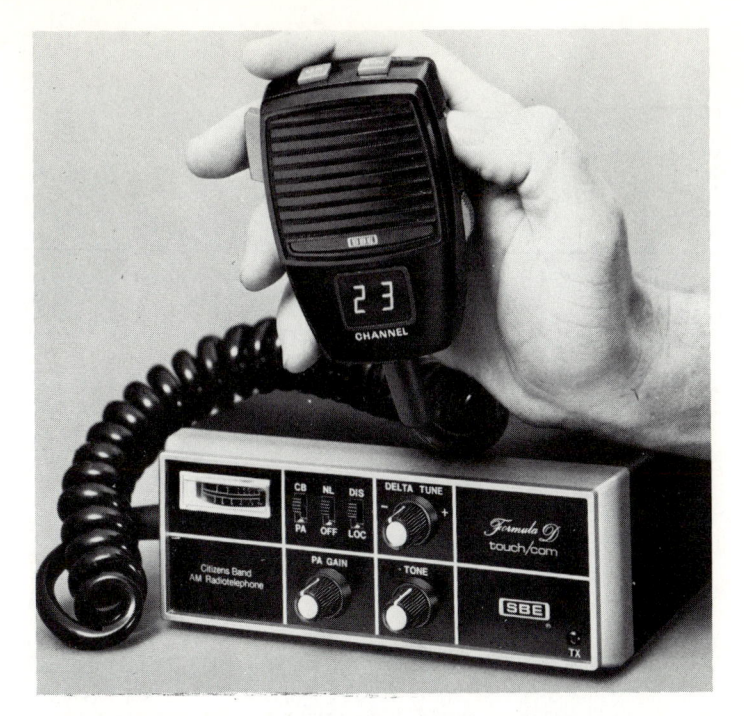

FIG. 5-9 Model SBE-32 CB transceiver. (Courtesy Linear Systems, Inc.)

Crystal Oscillator. A large part of the transceiver consists of oscillating, frequency converting, and RF amplifying stages.

Just about every oscillator used in a CB radio is based on the Colpitts circuit in Fig. 5-10. Feedback exists through the crystal and capacitors C1 and C2 from the collector to the base of the transistor. The amount of voltage fed back is determined by the ratio of the two capacitors, C1 to C2. In fact, one condition that must be met if the circuit is to oscillate is for the beta of the transistor at the oscillating frequency to be greater than C2/C1. Since this is a feedback oscillator, one other important condition must be satisfied if this circuit is to operate in the desired mode. The product of the voltage gain of the amplifier stage and the portion of the voltage fed back from the output of the transistor to the input circuit must be equal to or greater than 1. If this latter condition is satisfied, the gain of the circuit is infinite and oscillation occurs.

The crystal is a basic component in any stable oscillator circuit. Crystal may be thought of as L–C (inductor–capacitor) parallel resonant circuits. The schematic representation of a crystal and its equivalent circuit are in Fig. 5-11. Component values in the equivalent circuit are determined from the way the crystal has been cut, the size of the material, and its characteristics. Crystals are used as resonant circuits in many

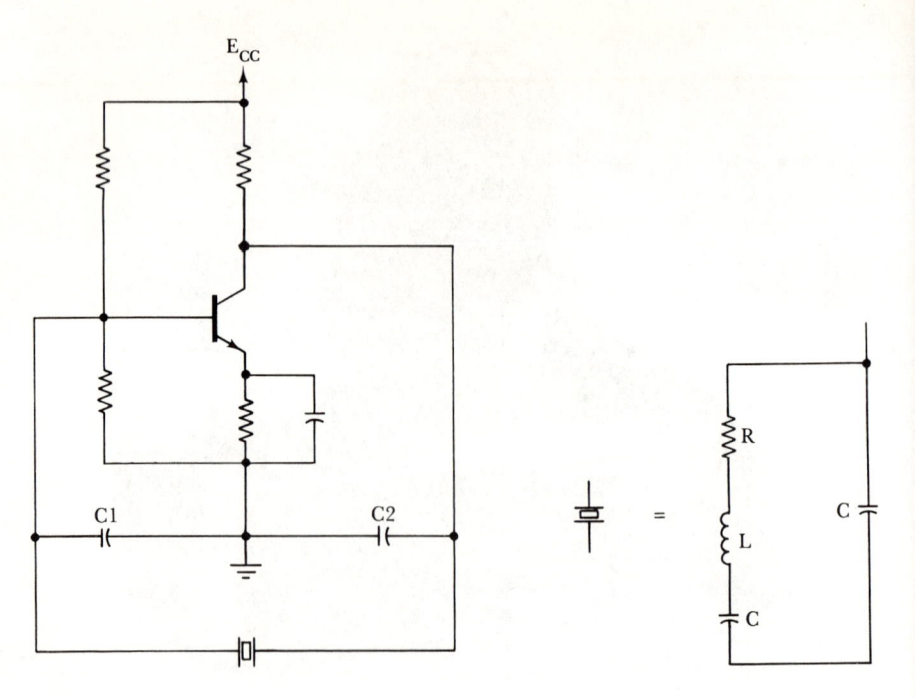

FIG. 5-10 Basic crystal-controlled Colpitts oscillator circuit.

FIG. 5-11 Equivalent circuit of a crystal.

different types of oscillators. In this capacity, the fundamental oscillating frequency is set by the L's and C's in the crystal, as well as by any capacitors and inductors in shunt or in series with the component.

Two practical oscillator circuits commonly found in CB radios using crystals in different arrangements are shown in Fig. 5-12. If you consider the capacities of the transistor junctions, you will realize that both circuits are basically Colpitts oscillators. Whereas the crystal in the Colpitts oscillator in Fig. 5-12a is used as a resonant circuit from the base to the emitter of the transistor, the crystal in the Pierce version of the Colpitts oscillator in Fig. 5-12b is in the actual feedback loop from the collector to the base. Many variations of these circuits are used by the various manufacturers, but they are all basically similar to those shown here.

In the Pierce circuit, the crystal is in a very vulnerable position. When replacement is required, an exact duplicate of the original component must be used if the oscillating frequency is to be maintained as required and the crystal itself is not to disintegrate.

Tuning of an oscillator circuit can be very simple if several factors are carefully noted. If you have tuning information from the manufac-

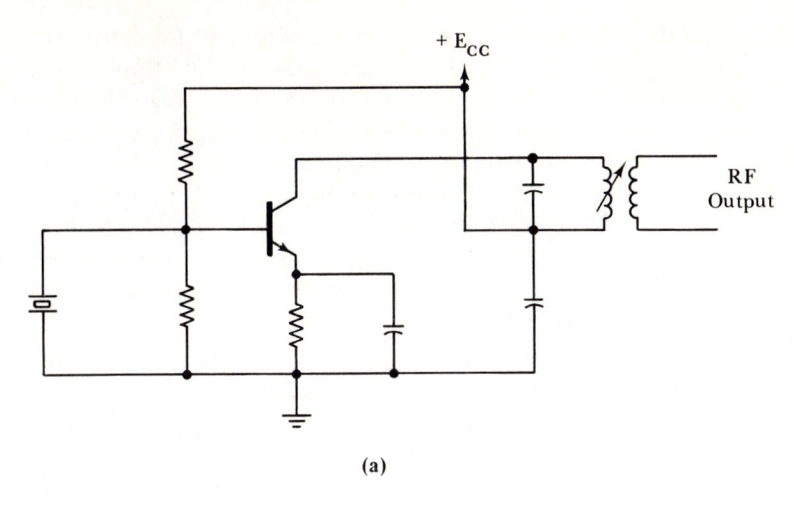

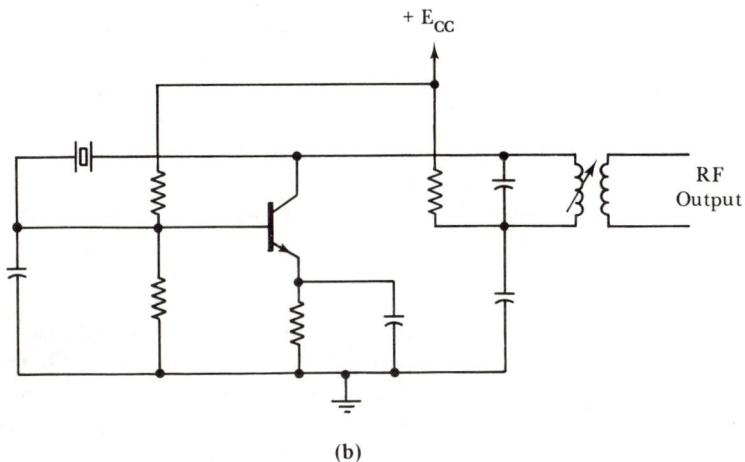

FIG. 5-12 Practical crystal-controlled oscillator circuits. (a) Colpitts, (b) Pierce.

turer of the equipment, follow his directions explicitly. If the manufacturer does not supply you with a procedure, then note the following three points:

1. Most oscillators operating at frequencies other than in the 30-MHz region supply signals that are at the fundamental frequency of the crystal. Above 30 MHz, frequencies are usually at the crystal's third overtone.
2. Oscillators do not work well if the tank circuit in the collector is peaked to the oscillating frequency. When tuning the tank cir-

cuit through a peak, you can note that the output drops rapidly on one side of the peak and gradually on the opposite side of the peak. You should adjust the tank so that resonance is at a point on the gentler slope. Voltage output from the oscillator should be 80 to 90 percent of that at the peak. Once adjusted, you should key the microphone on and off to be sure that the circuit oscillates each time the switch is thrown to transmit. Also turn the transceiver on and off several times to be certain that oscillation commences the instant the set is turned on. Operation should not be intermittent or erratic.

3. You cannot determine if you are properly performing the adjustment in point 2 without using point 3. Do not excessively load the tank circuit when making adjustments. Use a meter or oscilloscope with an RF probe. Connect the RF probe to the secondary winding of the tank circuit transformer to measure relative RF voltages. When checking frequency with a digital counter, it too must present a high impedance to the circuit under test. The counter may be connected to the secondary winding of the tank transformer or, better yet, to the output of the transceiver. If you connect it to the transformer, be sure the counter does not load it excessively. Use an RF probe between the transformer and counter for isolation. Approximate frequency checks can be made by loosely coupling the coil of a dip meter to the coil in the tank circuit.

An initial check of a transceiver involves tests to determine if the frequencies are proper at the various stages following the oscillator as well as at the output of the oscillator itself. In any of the systems described above, determine if the output frequency from the oscillator proper or from the PLL is as specified for the particular transceiver that you are troubleshooting. Then proceed with checking frequencies in other parts of the circuit. A digital counter can be used to let you accurately determine the frequencies of oscillation and conversion. A good frequency approximation can be made from the display on a wideband triggered sweep oscilloscope by noting the time of one cycle in milliseconds or microseconds, dividing 1 by the time for 1 cycle, and multiplying the quotient by 1000 if the time read from the scope calibration is in milliseconds, and by 1,000,000 if the reading is in microseconds.

When checking frequency, start the tests in the receiver section of the transceiver. Different frequencies can be found in this circuit. IF signals are present only when there is RF at the antenna. RF must be at the frequency of the channel used when making tests. It may be available from a CBer transmitting a message, from a second transceiver in your

Amplitude-Modulation Transmitter Blocks

service shop, or from the output of a signal generator loosely coupled to the antenna of the CB radio being tested. RF supplied must be precisely equal to the one assigned to the particular channel involved in the test. If a signal generator is used to supply the RF, its frequency should be continuously monitored and accurately adjusted by using a digital counter in the operation.

Three frequencies can usually be found in the transmitter section of the transceiver. Two are the outputs from crystal-controlled oscillators and the third is the output from the mixer after the two oscillator frequencies have been combined. It is this last frequency that must be very accurate as it is amplified, modulated, and transmitted. All three frequencies must be checked on every channel by using a digital counter to assure that operation is within FCC requirements.

If no signal is transmitted, the first tendency is to suspect the oscillator. Do not overlook items outside the transceiver proper. Sometimes the lack of signal in either the receiver or transmitter section may be due to an oversight. Be sure to connect the microphone to the transceiver. The microphone must be connected to any CB radio using electronic switching if the transceiver is to be operative in both the transmit and receive modes. As for transceivers using relay switching, the microphone may not have to be connected to the CB radio for receiving signals, but it definitely must be connected when transmitting. In either type of setup, do not forget to key the switch on the microphone when checking frequencies in the transmitter section. If no signal is supplied by the oscillator, be sure to check the switch in the microphone, as well as the microphone cable, before accusing the transceiver of any wrongdoings.

Once all outside factors have been accounted for, tests must be made on the transceiver to determine where the defect is. Only the oscillator, mixer stage, or power supply can be at fault. Here we shall assume that you have found the correct supply voltages at the mixer and oscillator stages, but that there is no output from the oscillator. You can assume that nothing is wrong with a mixer before you get the oscillator to oscillate at the proper oscillating frequency.

You can check the overall circuit by substituting RF from a generator for the suspected oscillator in the CB radio. Note where the output from the suspected oscillator is connected to the rest of the circuit. Using a small capacitor (under 5 pF), connect the RF generator to that point. Monitor the frequency of the RF generator with a digital counter so that you are certain that the proper frequency is injected into the circuit. Vary the amplitude (or voltage) of the output signal from the generator. If operation is restored to normal at any one setting of the output control of the generator, you can be sure that either the suspected oscillator is not operating properly or that the coupling circuit is at fault.

If the oscillator circuit does not work, the first thing you would normally suspect as defective is the crystal. Although this component is extremely vulnerable, other items that may actually be defective could make it appear that the crystal is the culprit. It is wise to check the transformer in the collector circuit as well as the capacitor across that transformer before jumping to the conclusion that the crystal is bad. The inexpensive capacitor in the tank circuit may be shorted, so do not replace the crystal before reassuring yourself that all components in this portion of the circuit are in good order. You can check the crystal by switching channels. If the oscillator works properly on one channel and not on another, you can be pretty sure that it is a problem with the crystal, unless, of course, it is the switch.

The switch can be a tricky component when it comes to RF circuits. Corroded contacts can make the circuit completely inoperative. The connection between the wiper and the stator may be poor or nonexistent. Corrosion should be removed by applying TV tuner contact cleaner to the lugs and wiper.

Solder flux between stator lugs on the switch can form an electrical path for shunting one crystal across another. Two crystals are thereby connected to the oscillator transistor at the same time. The frequency supplied by the oscillator differs from what would be present if only the selected crystal were in the circuit. Flux between two lugs can even make it appear that a dead crystal is operating, although off frequency. This can occur when a crystal on an adjacent set of switch contacts is placed into the oscillator circuit by solder flux. Flux must be thoroughly cleaned off the switch wafer before continuing with troubleshooting procedures.

Another important item that must be carefully scrutinized before concluding that the crystal is defective is the solder connections. Make sure that all connections are good. There must be no cold solder joints. Good soldered connections are smooth and shiny; cold solder connections have a pitted or grainy look. Also be sure that all flux has been cleaned off the board and switch. You can use any nontoxic cleaner for this purpose.

Along with the tank circuit, switch, and solder connections, it is wise to check that the transistor is in proper working order. It can be tested while connected in the circuit by using one of the testers described in Chapter 4.

After going through all the steps to prove that the expensive crystal is in good shape, you may finally realize that the worst has happened. The crystal is the only item that can be defective. But all is not lost even now. Once in a while, the crystal can be revived by iron-to-crystal resuscitation. Hold a hot soldering iron to the pins or solder lugs of the crystal until its case gets extremely hot. Without moving the crystal, remove the

Amplitude-Modulation Transmitter Blocks

iron. Let the crystal cool. Replace it into its circuit. If the circuit works, you have revived the crystal. Check to be sure that the oscillator is operating at the proper frequency when using the revived crystal.

Many of the reasons that problems arise in CB radios have already been discussed. Symptoms due to these problems are useful indicators in guiding you to the actual defect.

It was noted that flux between stator lugs on a switch can supply a path for a crystal wired to a nearby lug. It was indicated that the neighboring crystal can either take over for a defective one or operate in conjunction with a crystal that has been selected because of the switch setting. Symptoms due to this type of defect are (1) erratic operation or (2) in the same switch setting, one channel may be received while a different channel is being transmitted.

Intermittent reception and/or transmission on one channel or on several adjacent channels is an important symptom in guiding you to the defective component. This symptom may appear if the channel selector switch has corroded contacts. If *all* channels are erratic or even inoperative, a more likely cause than corroded contacts is a bad connection to the common terminal of the crystal bank. The same symptom could also indicate that the connection from the bank of crystals to the switch has not been soldered properly. You may, however, have to look further. Check all components in the transistor circuit along with the transistor itself. Be sure that the L–C tank circuit is tuned properly and not peaked. There will, of course, be no output from the transmitter section if there is no dc voltage at the oscillator transistor. In this case, check the relay or electronic switch circuit in the transceiver proper, as well as in the microphone.

If *several* channels are dead or intermittent (usually several adjacent channels), the channel selector switch is probably defective. (We are not picking on the switch, but mechanical components are the most likely to go bad in any piece of electronic gear.) Within this symptom category, it is also wise to check the wiring to the oscillator circuit. A defective crystal is a distinct possibility when there is a symptom of this type.

Let us finally assume your receiver works on all channels, but performance is only fair on some of them. Two things must be suspected immediately. One, and possibly more than one, of the crystals may not be on frequency. The second factor is an improperly tuned L–C tank. Although misalignment here affects all channels, one channel or a group of channels may suffer more than all the rest.

Mixers

Mixers are commonly associated with the receiver section of the CB radio, for it is here that the received signal combines with the output from the local oscillator. The desired IF is extracted from the various frequencies at the output of the mixer. In this section of the transceiver, the mixer is frequently referred to as the first detector.

In transmitters, mixers are used in the circuits for synthesizing the RF. All mixer circuits are nonlinear so that two or more signals can be combined. Semiconductors are nonlinear devices, as the ratio of the voltage to the current (or the resistance of the device) changes with the amount of current flowing through it. A simple diode can be used as a mixer. One arrangement of this type is shown in Fig. 5-13. Signals f1 and f2 are at different frequencies and are isolated from each other by resistors R1 and R2. After passing through the diode, either the sum or difference of the two frequencies is selected by T and passed on to the buffer stage.

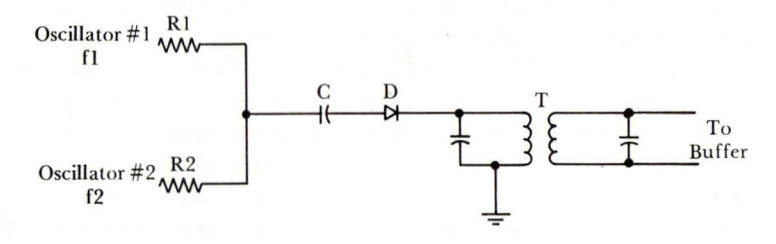

FIG. 5-13 Diode mixer.

Transistor mixer circuits commonly used in transmitter sections of CB radios are shown in Fig. 5-14. In Fig. 5-14a, the two signals are mixed in the base of Q1. The difference frequency is selected by the L–C circuits in the collector of the transistor and passed on to a predriver stage, Q2, for amplification. In Fig. 5-14b, one signal is coupled through a capacitor to the base of transistor Q14, while the second one is induced in the emitter circuit through transformer T7. As before, they are combined in the transistor and the difference frequency is selected by autotransformers T8 and T9 in the collector circuit. The actual mixer stage as used in a practical circuit is shown as an element involving Q14 of the Johnson 123A radio, Fig. 5-3.

FETs can also be used as mixers. A typical arrangement is shown in Fig. 5-15. Here both signals are applied to the gate, and the difference frequency is selected by tuning the transformer in the drain. The secondary of the transformer is usually connected to a predriver where the signal is amplified and sent on to the buffer stage.

Should there be no output from the mixer, measure voltages at the various elements of the transistor. Be sure the switch on the microphone is keyed on to put the transceiver into the transmit mode. If there is no voltage at the collector (or drain), check the power supply and the decoupling circuit. (The decoupling circuit consists of resistors in series

Amplitude-Modulation Transmitter Blocks

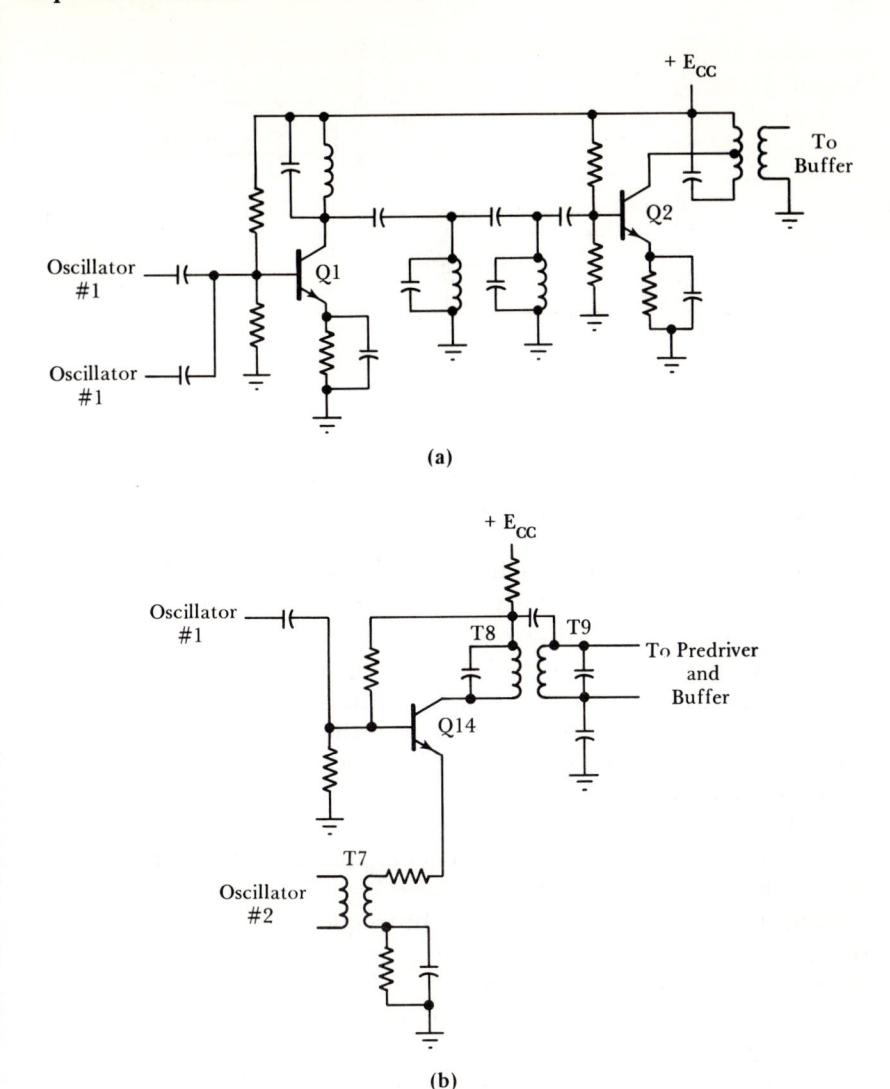

FIG. 5-14 NPN transistor mixer and predriver. (a) As used in the KRIS XL23, (b) As used in the Johnson 123A.

with voltage from the power source and all capacitors shunting these resistors to ground.) A resistor may have overheated, changed value, or opened. A capacitor in the circuit could easily have shorted, thus bypassing all voltage along the dc supply line to ground. Decoupling resistors can be destroyed by a shorted capacitor. If the short does not ruin the resistor, it can still keep voltage from reaching the collector or drain of the transistor.

An absence of voltage at the collector can also be due to an open coil in

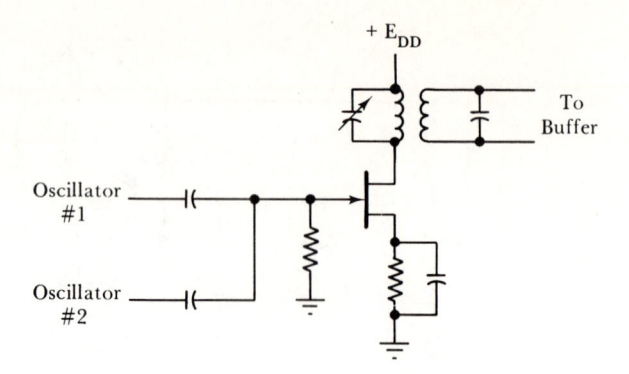

FIG. 5-15 FET mixer.

the primary of the coupling transformer in the collector circuit. Check for voltage on both sides of the coil before deciding that the problem is in the power supply network.

Check voltage across the base–emitter junction of the bipolar transistor. The importance of this voltage has been discussed in Chapter 4. Improper voltage can be due to a defective resistor in the base or emitter circuit, an open coil in the secondary winding of T7 in the circuit drawn in Fig. 5-14b, or a defective transistor. Check the transistor with an in-circuit transistor tester. Remove the transistor from the circuit if the test shows the device to be bad, and recheck it out of circuit to reassure yourself that it really is defective.

Be certain that all transformers and coils at the output of Q14 are tuned to the desired frequency or frequency band. It is best to use channel 12 for this purpose. Adjust all variable L–C circuits in the mixer and predriver circuits for a peak output at the frequency in question, unless the procedure given by the manufacturer of the radio states otherwise. Monitor the frequency using a counter and check the voltage using an RF probe and TVOM (or wideband oscilloscope) at the output of the predriver. Any L–C circuit that cannot be adjusted, however slightly, when tuned should be suspected of having a defective capacitor or inductor.

Buffer and Power Amplifiers

A buffer amplifier is placed between the oscillator and power amplifier to separate the two stages. By isolating the oscillator from the power amplifier, good frequency stability can be achieved.

Many different buffer amplifier and power output circuits are used in CB radios. They are all essentially embellished variations of the one shown in Fig. 5-16. Here buffer (or driver) transistor Q1 drives output device Q2. Both transistors are in a Class C mode of operation, which

Amplitude-Modulation Transmitter Blocks

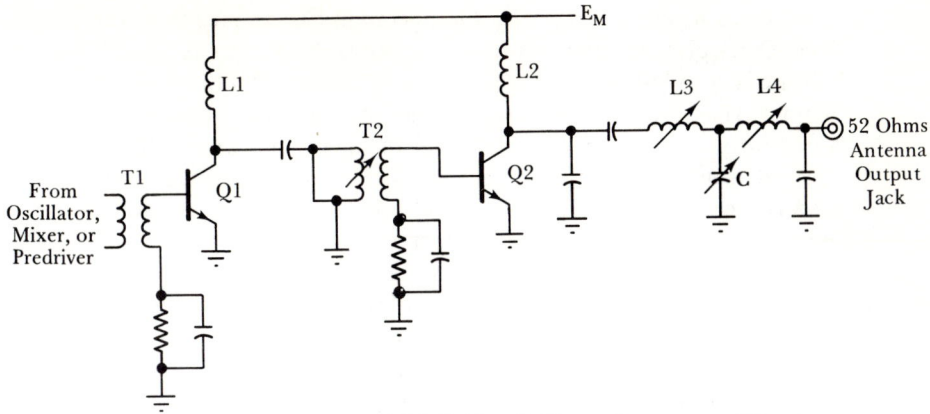

FIG. 5-16 Buffer and power amplifier stages.

indicates that they do not conduct current when there is no **RF fed** to the circuit from an earlier stage.

Unless directed otherwise by the manufacturer of the **transceiver**, use the following procedure to align the stages. Connect an **RF wattmeter** across the antenna jack on the CB radio. Use a 52-Ω resistor as a dummy load at the output jack of the wattmeter, instead of an antenna. Set a VOM to a range that can safely accommodate 500 mA and connect it from the collector of Q2 to ground. A variable regulated power supply set to 13.8 V should be connected from E_M to ground, but do not turn it on yet.

Now disconnect one lead from choke coil L2, so that there is no voltage at the collector of Q2. Turn on the power supply and adjust T2 for a peak reading on the VOM. Remove the VOM from the circuit after having adjusted T2 and connect it in series with L2, restoring that circuit and voltage to the collector of Q2. Now set variable coils L3 and L4 as well as capacitor C1 for a maximum reading (maximum output) on the RF wattmeter. Should the output power exceed 4 W, or should the milliammeter indicate more than 362 mA, L3 and L4 must be readjusted. Decrease the inductance of L4 and increase the inductance of L3 until the required readings are achieved. It is best to change the inductance of both components by similar amounts rather than radically upset the adjustment on just one of the coils. Readings indicated here as acceptable maximums are not only required if the transceiver is to operate within FCC regulations, but may also be necessary to safeguard the operation and prolong the life of the output transistor. Remove the VOM from the circuit and restore the original connections to L2.

Transceiver embellishments that may be at the power output device include lights to indicate that RF is present at the antenna circuit. In the Royce 1-602A, this bulb is connected through a transistor to the antenna terminal as shown in Fig. 5-17. Operating in Class C, no collector current flows through the transistor unless signal is applied to its base. When RF voltage is at the antenna output jack, it is also at the base of the transistor. When turned on, current flows through the transistor and bulb, lighting it. As the output power from the transceiver varies somewhat with the percent of audio modulation, the intensity of illumination of the bulb is affected by the amount of modulation present. The light thus flickers somewhat as you modulate the RF by speaking into the microphone.

A meter circuit can be used instead of the bulb to indicate relative RF output power, as well as the presence of modulation. It too is connected at the antenna output jack. A circuit employing a meter, as used in the Hy-Gain 670A, is shown in Fig. 5-18. RF is rectified by the diode, filtered by shunting capacitor C, and applied to the meter movement M through the variable control R. Meter sensitivity is adjusted to the manufacturer's specifications using R. If the meter does not read, does not read properly, or is erratic, suspect the diode, capacitor C, and the meter movement M, in that sequence. Variable control R can go bad if sensitivity adjustments are made frequently.

The RF output transistor can be destroyed if there is no antenna connected at the output jack. An embellishment found on some transceivers is the antenna warning light. The bulb lights when an antenna has *not*

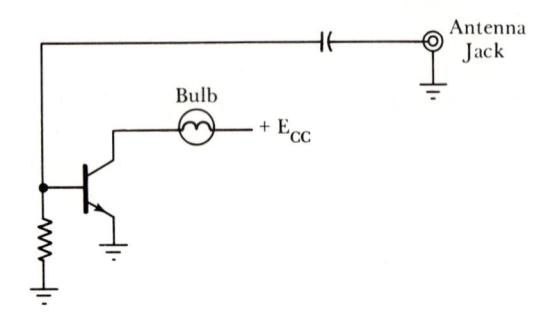

FIG. 5-17   Bulb indicates if RF and modulation are present at the output.

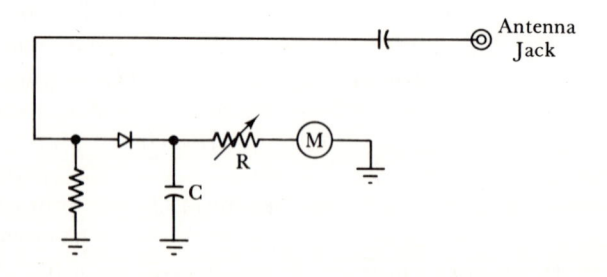

FIG. 5-18   Meter deflection indicates if RF and modulation are at the output.

Amplitude-Modulation Transmitter Blocks 113

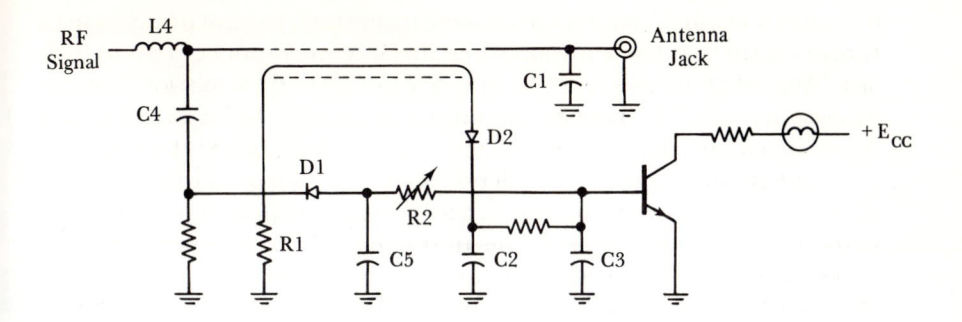

FIG. 5-19 Bulb lights when the antenna is *not* connected at the jack.

been properly connected to the jack. It is a warning to turn off the transmitter and connect the antenna.

A circuit used for this purpose in the Midland 13-882B is shown in Fig. 5-19. Noting the output circuit in Fig. 5-16, the outer shield of a short length of coaxial transmission line cable is wired between L4 and the antenna jack. When RF voltage is at L4, current flows through the shield, for the current path is complete through C1. Because of C1, this current flows regardless of whether an antenna is or is not connected at the jack. Current is induced into the inner conductor because of the current flowing in the shield. The current in the inner conductor is rectified by D2, filtered by C2 and C3, and applied, as shown, to the base of the transistor. Because the voltage developed here is positive with respect to ground and emitter, the transistor is turned on. Base current, multiplied by the beta of the transistor, is in the collector circuit. The bulb is turned on by this current.

Next, consider the voltage at L4. When an antenna or resistor load *is* at the antenna jack, the voltage at L4 is relatively high. It is fed to D1 through C4, rectified by D1, filtered by C5 and C3, and applied to the base of the transistor. This voltage is negative with respect to ground. It bucks the voltage at the base owing to current through D2. R2 is adjusted so that the voltage from D1 is sufficient to offset the voltage from D2 at the base and turn off the transistor. Under this condition, the bulb does not light.

Voltage at L4 is low when there is no load at the antenna jack. It is insufficient to cancel the positive voltage from D2 at the base. Consequently, the transistor is turned on by the positive voltage, and stays on. The bulb is lit, indicating that the antenna is not connected to its jack.

Should the circuit fail, the semiconductor and bulb are the most likely components to go bad. After numerous adjustments, control R2 may open. Be sure also to check all solder connections, as cold solder joints, especially to the shield in the cable, can upset the overall performance of the transceiver as well as the operation of this indicating circuit.

As for the driver and RF power output circuits proper, if there is no output power, check all voltages at these transistors. Output devices can be shorted if there is transmission when the antenna is not connected to its jack. A short at the jack, possibly due to a defective transmission line, can likewise irreparably damage the output transistor. Before removing it from the circuit, use a transistor tester to check if a short does exist.

Other components that will not permit RF to pass to the output are open series inductor or shunting capacitors feeding the antenna jack. Also suspect these components if you find it impossible to tune a variable inductor or variable capacitor in one of these circuits. Not only could the component that you are attempting to adjust be defective, but the associated mate (capacitor or inductor) in its resonant circuit may have opened, shorted, or changed value.

Modulator

The audio section of the transceiver serves two functions. When receiving, the minute audio signal from the second detector is amplified here and fed to a loudspeaker. This loudspeaker is disconnected from the circuit when the transceiver is set in the transmit mode of operation. Instead, the output from the amplifier is applied to the driver and RF power amplifier stages to modulate the carrier. When transmitting, the input to the amplifier is no longer from the second detector in the receiver section of the CB radio, but is derived instead from a microphone and a special microphone amplifier stage of voltage gain. A typical circuit is shown in Fig. 5-20.

In the receiver (REC) setting of the switch in the microphone, the detected signal is amplified and fed through capacitor C1 and resistor R1 to the base of Q2. When its switch is keyed for transmitting, the microphone supplies audio to be amplified by Q1. The amplified signal is applied to the base of Q2 instead of the detected signal from the receiver stages.

One REC–TRANS switch section is shown connected in the lead from the receiver section to Q2, and a second switch section is shown between the microphone and Q1. Switching is usually not done in these circuits. The microphone and receiver output are connected permanently in the circuit. Instead of the switching as shown, power is removed from the microphone amplifier circuit when receiving and from the receiver electronics when transmitting.

Regardless of the switching arrangement, signal at the base of Q2 is amplified by that transistor. The load in the collector of Q2 is the primary of driver transformer T1. Properly phased signals are applied to the bases of Q3 and Q4 owing to the characteristics of the center-tapped secondary winding of T1.

Transistors Q3 and Q4 form a push–pull pair. In this type of circuit, neither transistor is conducting (they are cut off) until there is signal

Amplitude-Modulation Transmitter Blocks

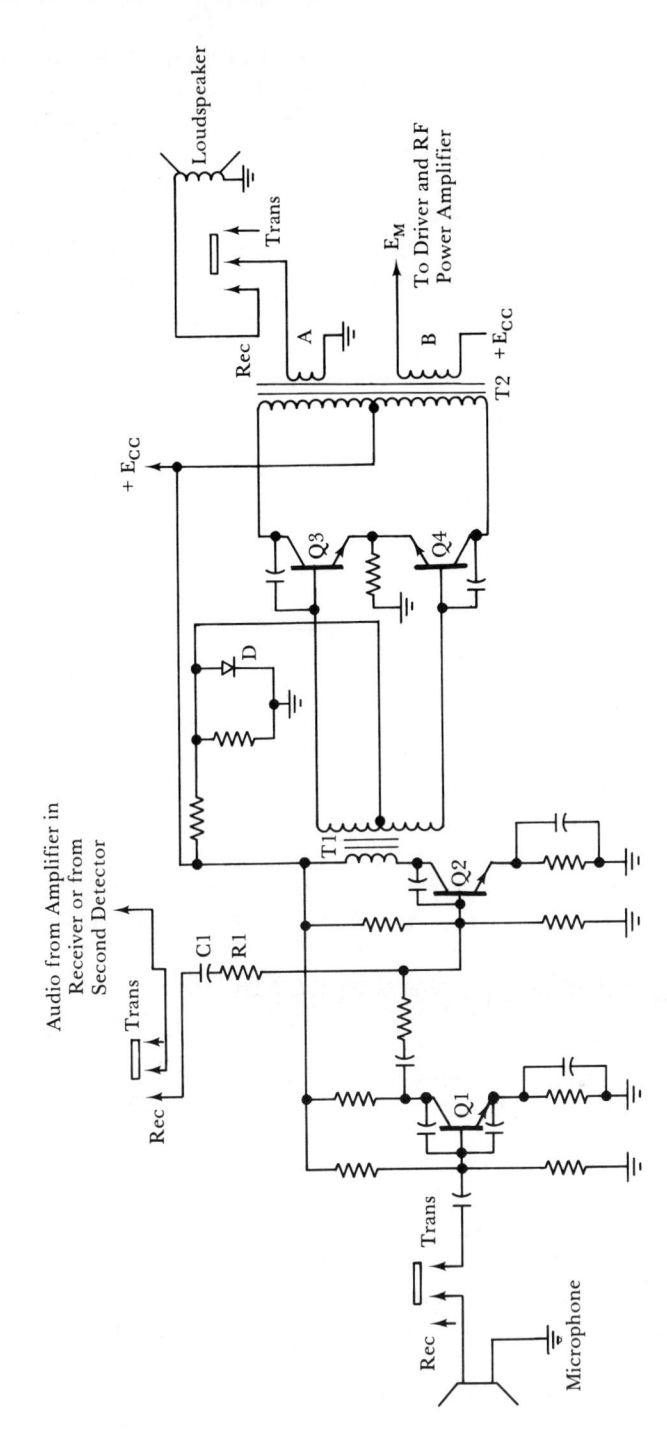

FIG. 5-20 Modulator. Slide switch sections are physically in the microphone.

of the proper polarity applied at their bases. During the half of the cycle when the end of the transformer connected to the base of Q3 is positive with respect to its center tap, the base of the transistor is positive with respect to its emitter. Q3 conducts. During the alternate half of the cycle, the end of the transformer connected to the base of Q4 is positive with respect to the center tap. Q4 conducts during this half of the cycle. The push–pull circuit requires that signal at the base of Q3 be 180° out of phase with signal at the base of Q4. The out-of-phase signals are amplified by these transistors. Each half of the primary winding of the output transformer, T2, is in the collector of one of the output transistors. After amplification, each out-of-phase signal appears across one half of the primary winding. The two out-of-phase components combine in the secondary winding to form an amplified version of the original signal.

It was just noted that each output transistor is cut off until there is signal at its base. But as you recall, there is about 0.7 V across a base–emitter junction (or across any silicon diode) when it conducts. There is no current through the junction, or through the collector circuit of the transistor, until there is at least 0.7 V across the junction. If you rely only upon the signal to turn on Q3 and Q4 from a 0-V level, the first 0.7 V of the signal is lost in each half of the cycle. This causes *crossover distortion*. To minimize this condition, a fixed 0.7-V dc supply is applied to the bases of Q3 and Q4 through the center tap of the secondary on T1. This positive voltage is across conducting silicon diode D, for it is forward biased.

To return to the recombined ac signal, it is induced into the two secondary windings of the output transformer. Winding A is connected to the speaker when the CB radio is set in the receive mode. In the transmit mode, one end of winding B is connected to the power supply E_{CC}. The other end is the voltage E_M used as the supply for the collectors of transistors Q1 and Q2 in the driver and RF power output stages in Fig. 5-16. Collector voltage supplied through winding B is not a fixed or solid E_{CC}. It varies with the audio signal in the winding. It is this voltage variation that produces a variable RF amplitude in the loads at the collectors of Q1 and Q2 in Fig. 5-16. RF amplitude varies with the audio signal. This amplitude variation is the modulated RF due to electronic peaks that began as speech at the microphone.

A signal tracer or oscilloscope can be used to locate a defective audio stage. Feed an audio signal at about 1000 Hz (any frequency from 400 to 2000 Hz will do) into the microphone amplifier. Keep the microphone connected to the circuit so that you can use it to trigger the relay or electronic switching circuit into the transmit mode. Be sure the input signal is very low, usually less than 2 mV, so as not to overload any of the stages. Then check for output signals at the various bases and

Amplitude-Modulation Transmitter Blocks

collectors of the transistors, starting at Q1 and proceeding through the circuit to Q3 and Q4. Voltages of increasing amplitude should be at each succeeding stage. Similar amounts of audio should be at both ends of coupling capacitors. Also check for audio across the various transformer windings. Audio voltages must be identical across both halves of a center-tapped winding.

One difficulty that may arise with the signal-tracing procedure when in the transmit mode of operation is that RF may leak into the audio circuit, masking the audio signal. To avoid this problem, it is best to troubleshoot the stages when the transceiver is in the receive mode. Feed audio to the input side of C1 rather than to the base of Q1, and proceed with the signal-tracing procedure as described. Once the defective stage has been located, voltage checks can be used as a guide to help you locate the faulty component. Should you suspect a transistor as the culprit, make tests on it using one of the in-circuit testers. Should all stages check properly, connect the oscilloscope across a secondary winding of the output transformer, and put the transceiver into the transmit mode. Whistle into the microphone. If there is no modulating voltage, even though it may be mixed with RF, suspect the microphone or microphone amplifier, as the rest of the audio circuit has already been checked.

Lack of modulation can also be due to an absence of voltage from the power supply or to an open or shorted component in the circuit. If all transistors are good, the next two troublemakers in order of likelihood are capacitors and transformers. Capacitors from the collector to the base and from the base to the emitter are in the circuit to keep RF out of the audio section. But if one of these components shorts, it will also keep audio out of the circuit. A transistor in a circuit with a shorted capacitor will appear defective when it is tested in circuit and will check good when it is tested as an independent component.

Distortion will run rampant if diode D opens or if there is a shorted winding in one of the transformers. Although an ohmmeter can be used to measure if a transformer winding is open, the resistance of a winding can seldom be used as an indication of whether there is or is not a shorted turn. Substituting a good transformer for the defective one is just about the quickest method of determining if the suspected component is really bad.

Q2 may be damaged owing to a shorted winding in transformer T1, whereas Q3, Q4, or both may become defective if there is a shorted winding in transformer T2. Considering this, always be careful not to accidentally short any transformer winding when troubleshooting the audio circuits.

Finally, you may find modulation is low or less than about 90 percent. Assuming that all limiter circuits are operating properly, this condi-

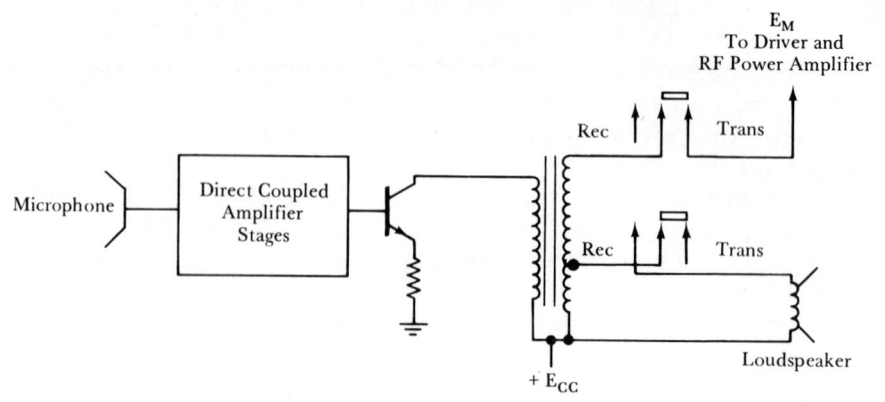

FIG. 5-21 Direct coupled modulator.

tion can be due to a defective microphone cartridge or shorted diode D. Be sure to check that there is about 0.7 V across this diode at all times.

Direct-coupled circuits (see Fig. 5-21) are sometimes used in the audio section. The output device in this case is usually a single transistor rather than the push–pull pair previously described. Even though the electronic circuit consists of direct-coupled stages without the intervening coupling capacitors, a transformer is used in the collector circuit of the output transistor. A loudspeaker is connected from the common ($+E_{CC}$ in this circuit) to a tap on the secondary winding of the transformer, to match the loudspeaker to the circuit. $+E_{CC}$ is used as the common point for both the loudspeaker and the transformer. If the loudspeaker were connected to ground instead of to $+E_{CC}$, dc current would flow through its winding. This is undesirable.

Direct current does flow through the entire secondary winding of the transformer when transmitting, for E_M (see Fig. 5-16) is at the upper end of the winding to supply the modulating voltage. In this mode of operation, the loudspeaker is disconnected from the circuit.

Modulation Limiter. FCC rules state that a modulation limiter must be an integral part of every transmitter circuit. Its function is to limit the size of the audio signal to levels capable of modulating the RF to just under 100 percent. A typical circuit accomplishing this is shown in Fig. 5-22.

The audio portion of the modulating voltage is fed back through C1, rectified by D1, and filtered by capacitors C2 and C3. Positive dc voltage with respect to ground is across C3 because of the rectified and filtered audio. Applied to the emitter circuit of the microphone amplifier transistor, or across R2, the direct current affects the bias and

Single-Sideband Transmitter Blocks

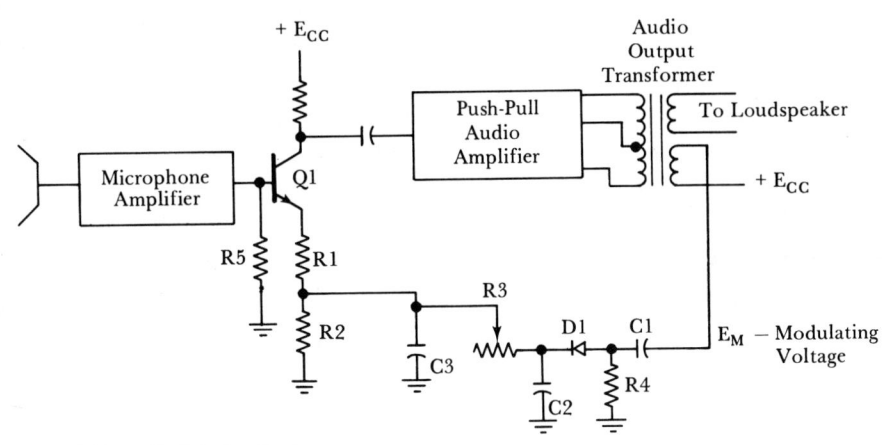

FIG. 5-22 Modulation limiter.

hence the gain of the transistor. Gain is reduced as the voltage across R2 is made more positive with respect to ground.

In any *npn* transistor, gain is reduced when the forward-bias voltage across the base–emitter junction is reduced. In the modulation limiter circuit, positive voltage from R2 is applied across the junction through resistor R5. It tends to reduce the forward-bias voltage that has been previously established across the junction by other circuitry. As the audio modulating voltage is increased, positive voltage across R2 rises to limit the overall gain to predetermined levels. R3 is adjusted so that the correct voltage is applied across R2 to do the job properly.

Adjustments can be made by observing the modulation pattern on an oscilloscope connected in the antenna circuit. Feed a large audio signal to the microphone amplifier. Set R3 so that modulation is limited to 100 percent (see Fig. 2-8c). If the modulated waveshape is not sinusoidal, reduce the audio output from the generator to levels that are capable of overmodulating the RF, but will still provide you with a sinusoidal display when R3 is adjusted to one of its extreme settings.

As in all circuits, the components most likely to fail are diode D1 and the electrolytic capacitors.

Single-Sideband Transmitter Blocks

At this juncture, we suggest that you turn back to Chapter 2 and refresh you memory on SSB. Instruments to measure SSB power were discussed in Chapter 4. Crystal-controlled oscillators, synthesizers, and the PLL circuits used in SSB transceivers are similar to those described above for generating RF for the AM type of broadcasting.

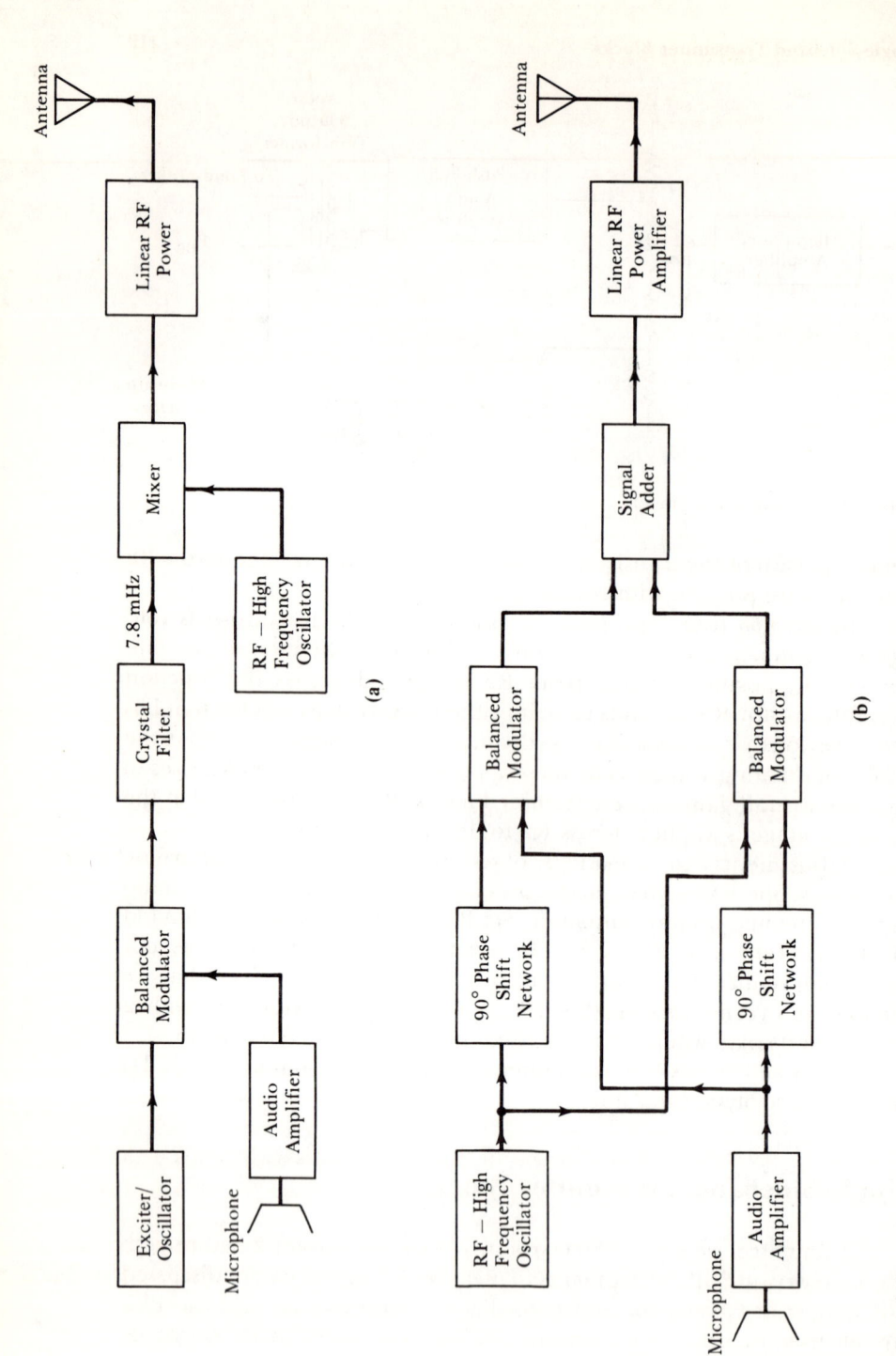

FIG. 5-23 Two systems for producing SSB signals. (a) Filter method. (b) Phasing method.

Single-Sideband Transmitter Blocks

Basically, two systems are used for producing SSB signals. Block diagrams of both are shown in Fig. 5-23. The SSB system using the phasing method shown in Fig. 5-23b is seldom, if ever, applied to CB radios because the circuit is complex and adjustments are critical. It is described here, however briefly, along with the commonly used filter method in Fig. 5-23a, because it may occur in your work.

In the circuit using the phasing method, the carrier frequency is generated in the RF HF oscillator. It is split into two identical signals, with the only difference between them being that the phase of one has been shifted by 90° with respect to the other. Similarly, a phase-shifting network is used for supplying two audio signals differing by only a 90° phase relationship. Two balanced modulators are in the circuit. Each combines one audio and one RF signal to provide the upper *and* lower sidebands at its output, while suppressing the carrier. Both sets of sidebands from the two balanced modulators are combined in the signal adder, so that one sideband is canceled while the other is increased in amplitude. The generated SSB signal is amplified, ideally without distortion, in the linear RF amplifier before being fed to the antenna for radiation into the air.

In the detailed discussion to follow of the filter method of SSB signal generation, several oscillator frequencies are indicated. Although not the only frequencies that can be used in the various circuits, they seem to be the ones preferred by most manufacturers.

RF output from a crystal-controlled exciter oscillator and signal from the audio amplifier circuit are combined in a balanced modulator. Two oscillator frequencies can be generated in the exciter stage. When the transceiver mode switch is set to the lower sideband (LSB) position, oscillation is at 7.8015 MHz; in the upper sideband (USB) position, a signal at 7.7985 MHz is generated. Whatever signal is used, it is combined with the audio in the balanced modulator. Two sidebands, the LSB and the USB, are at the output of the balanced modulator while the "carrier" (7.7985 MHz or 7.8015 MHz) is suppressed.

In CB radio applications, audio consists of frequencies up to about 3000 Hz. The middle of this narrow audio band is 1500 Hz or 0.0015 MHz. When using the LSB carrier, 7.8015 MHz, two sidebands are formed in the balanced modulator. If the audio is 0.0015 MHz, one sideband is at 7.8015 + 0.0015 MHz, or 7.8030 MHz, and the other sideband is at 7.8015 − 0.0015 MHz or 7.8 MHz. The lower of the two sidebands is selected by the filter in this case, while the 7.8030-MHz sideband is rejected. Of course, the filter will not only pass 7.8 MHz, but will also pass a range of frequencies from 7.798500 to 7.801500 MHz (or 7.8 MHz + 1500 Hz and 7.8 MHz − 1500 Hz), so that the entire required portion of the audio band from 0 to 3000 Hz (0.003 MHz) will be produced. The 7.8 MHz is only the center frequency of the filter, just as

0.0015 MHz is the center frequency of the audio band as applied to CB broadcasting.

Similarly, when the transmission type (or mode) switch is set to USB, the 0.0015-MHz audio signal combines with the 7.7985-MHz exciting frequency. Now the sum frequency, 0.0015 + 7.7985 MHz, or 7.8 MHz, is selected by the filter.

You may justifiably question why two exciter frequencies are required, one for the USB and one for the LSB. It becomes obvious when you realize that one filter (normally a crystal filter) is used regardless of the sideband to be transmitted. If 7.8 MHz is to be the center of the LSB band, the suppressed RF carrier must be above 7.8 MHz. Similarly, the suppressed RF carrier must be below 7.8 MHz if it is to be the center frequency of the USB.

The 7.8 MHz from the filter is mixed with RF from an HF oscillator, producing SSB signals from the combination at the proper frequency in the 27-MHz band for transmission. Magnified by the linear power amplifier, RF pulses characteristic of SSB waveforms are fed to the antenna system.

Most transceivers designed primarily for SSB broadcasting are also capable of transmitting standard AM signals. When the mode switch on the front panel of the CB radio is set to the AM position, the balanced modulator circuit and crystal filter are bypassed. Signal from the 7.7985-MHz oscillator is combined with the output from the HF oscillator, AM modulated, and fed to the RF power amplifier and antenna. The circuit effectively reverts to one of those discussed earlier in this chapter.

Several of the blocks in Fig. 5-23 differ in detail from similarly labeled blocks shown in the drawings of AM systems. Other blocks encompass circuits and concepts not needed in CB radios using only standard AM transmitting and receiving methods. Here we shall describe each block peculiar to SSB broadcasting and its characteristics.

Exciter–Oscillator

Colpitts crystal oscillator circuits, described earlier, are used here in the exciter circuit. An applicable crystal is selected using the three-position mode switch on the front panel. The switch is set to either USB, LSB, or AM. Defects that can develop in the oscillator circuits and troubleshooting procedures have been discussed above.

Audio Amplifier

Neither power nor high voltage is required from the audio amplifier for the balanced modulator. In the SSB transmitter, the audio output from the microphone is amplified by two or three stages of voltage gain and is

Single-Sideband Transmitter Blocks

then applied to a balanced modulator circuit. Excessively high audio voltage is undesirable here for it can produce spurious signals throughout the RF spectrum.

Balanced Modulator

Less than 0.7 V peak of audio is applied to terminals 2 and 4 of the balanced ring modulator circuit in Fig. 5-24. This audio voltage is too small to turn on any of the diodes. RF voltage from the secondary of transformer T1 is between the slider of control R and RF ground terminals 2 and 4. (Note that while terminal 4 is ground for both audio and RF, capacitor C1 is chosen to be an open circuit for audio and a very low impedance path for RF signals.) RF amplitude from T1 is almost ten times the amplitude of the audio from the transistor. As several volts of RF are applied to the modulator, diodes are turned on by this signal. During the portion of the RF cycles when the wiper of R is positive with respect to ground, diodes D1 and D3 are turned on. During the alternate half of the RF cycle, diodes D2 and D4 are turned on. Audio along with the RF passes through the diodes that are turned on, modulating the RF.

Signals from diodes D1 and D3, as well as signals from diodes D2 and D4, are across the output transformer T2. Each signal contains a carrier and two sidebands. The carrier signals cancel at T2 while only the two sidebands remain to be passed on to the crystal filter. Both R and C2 are adjusted for a minimum RF (or carrier) at the output of T2 when no

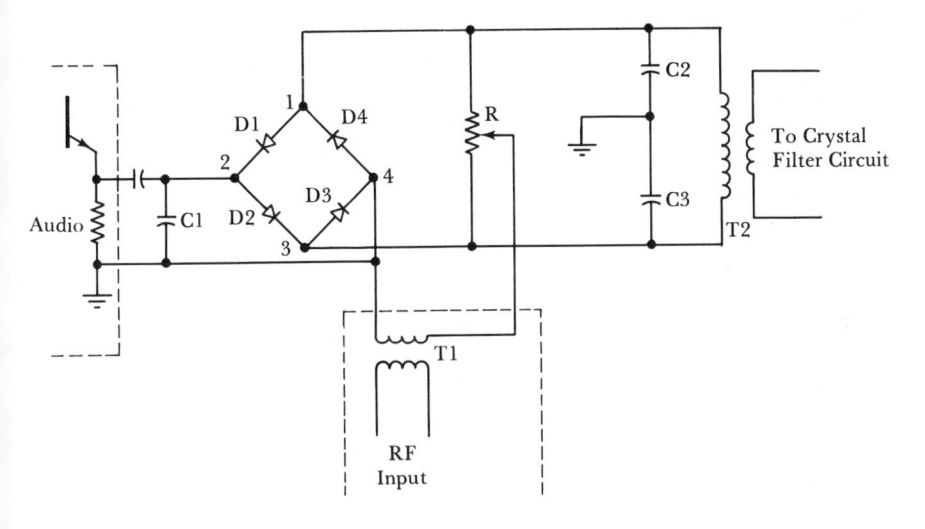

FIG. 5-24 Ring modulator.

audio signal is present. Adjust these variable components several times until you are certain that RF at the output transformer can no longer be reduced. Obviously, a minute amount of carrier will remain as no system is perfect.

If you cannot eliminate the bulk of carrier signal at the output of T2, check the variable components and capacitors C1 and C3. Diodes can, of course, break down and must be removed from the circuit for testing. You can check them with an ohmmeter, as discussed earlier, or on a transistor tester. When replacing a diode, be sure the replacement part is identical to the original. You can use an ohmmeter to check if the forward and reverse resistances of all diodes are the same, as they must be for best carrier rejection.

Crystal Filters

The equivalent circuit of a crystal used as the frequency-determining component in an oscillator is shown in Fig. 5-11. It is a high Q parallel resonant circuit. Crystals used as filters can also be represented by the equivalent circuit in Fig. 5-11. Because it is a parallel resonant circuit, it has an extremely high impedance at its resonant frequency. When the crystal is placed across a circuit, it shunts most signals to ground. Only RF at or very near the resonant frequency will not be affected. This narrow band of frequencies is allowed to pass from the circuit preceding the crystal to the circuit following it, without being attenuated. The crystal acts as a short circuit or very low impedance to all other frequencies.

One important characteristic of the crystal is its Q. High Q indicates that only a narrow band of frequencies will not be bypassed by the crystal. The band is just wide enough to accommodate one sideband. The crystal is a low impedance for RF at all other frequencies, including those in the second sideband. All frequencies may be in the circuit preceding the filter, but only the frequencies in one sideband survive the shunting by the filter. Because it has a narrow bandwidth and hence higher Q than can be achieved with L–C filters, only crystals are used in this application.

A piezoelectric slab is the basic material used in forming a monolithic crystal filter. Two pairs of metal electrodes are deposited on the slab, as shown in Fig. 5-25a. The equivalent circuit of the filter is shown in Fig. 5-25b. There are two parallel resonant circuits. One resonant circuit is formed by the input pair of electrodes; the second resonant circuit is formed by the output pair. An inductor is the electrical equivalent of the coupling between the two resonant circuits. Physically, signal is passed from one resonant circuit to the other by the mechanical vibrations in the crystal material. Because there are two resonant circuits,

Single-Sideband Transmitter Blocks

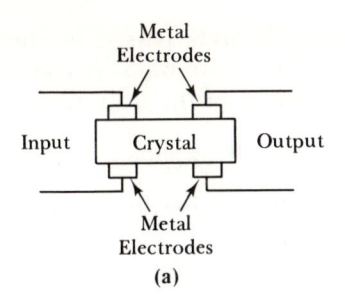

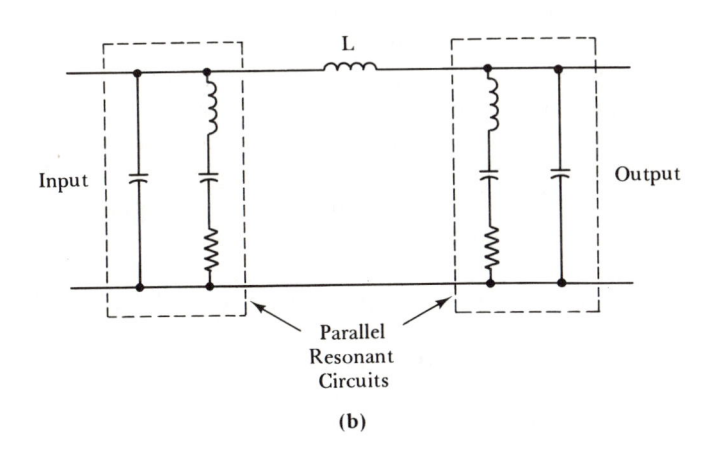

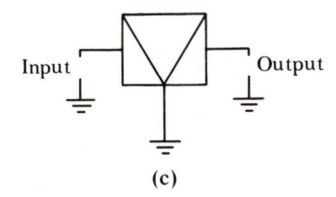

FIG. 5-25 Monolithic crystal filter: (a) physical construction, (b) electrical equivalent, (c) schematic representation.

the band passed by the filter is even narrower than the band passed by the simple filter first described. It can be narrowed further by adding more metal electrodes and hence more resonant circuits onto the slab. The schematic representation of the crystal filter is in Fig. 5-25c.

The band passed by the filter must be just sufficient to accommodate one sideband. If the band is too narrow, a resistor can be shunted across the crystal to reduce the Q.

In conventional SSB circuits, the output (secondary of transformer T2) from the balanced modulator in Fig. 5-24 is connected to the input side of the crystal filter. A transistor whose sole function is to amplify the narrow band around the 7.8 MHz passed by the crystal is normally connected to the output side of the filter. A transformer tuned to 7.8 MHz is the load in the collector of the transistor.

Radio-Frequency High-Frequency Oscillator

As in the synthesizer described earlier, two groups of oscillators are in the RF HF oscillator section of most 23-channel SSB transceivers. (Forty-channel SSB radios are normally built with PLL circuits.) There is a band of six crystals in one of the groups. Frequencies supplied here range from 11.705 to 11.955 MHz. In the second group, there is a choice of two oscillator circuits, each with a bank of four crystals. One of these circuits is activated when the mode switch on the front panel of the CB radio is set to LSB. In this instance, frequencies from 7.4585 to 7.4985 MHz are provided by the oscillator. The remaining oscillator circuit is activated in the USB setting of the mode switch. Here frequencies in the range of 7.4615 to 7.5015 MHz are generated. The output from the RF HF oscillator is a combination of a signal from the first group with a signal from the second.

Let us take as an example frequencies generated for the LSB and USB on channel 1. For the LSB, the frequency supplied by the synthesizer is 11.705 + 7.4585 = 19.1635 MHz; for the USB it is 11.705 + 7.4615 = 19.1665 MHz. These are obviously not the SSB frequencies that will be transmitted.

Mixer

Frequencies generated in the synthesizer of the RF HF oscillator must be added to the modulated 7.8 MHz from the crystal filter. In our example, the frequency to be transmitted on the LSB is 19.1635 + 7.8 = 26.9635 MHz; on the USB, it is 19.1665 + 7.8 = 26.9665 MHz. The 26.9635 and 26.9665 MHz are the two frequencies that are to be transmitted on channel 1 when using the LSB and USB modes of operation, respectively.

From Table 1-1, you can see that channel 1 has been assigned 26.965 MHz as its carrier frequency. This frequency is 1500 Hz (0.0015 MHz) above the signal supplied in the LSB mode, and 1500 Hz below that supplied for the USB mode. This is the desired goal, as the center frequency of each sideband is ideally 1500 Hz on either side of the assigned channel frequency.

Linear Amplifier

RF from the mixer must be amplified to provide the 12 W peak power (PEP) for application to the antenna. This is the job of the linear amplifier. It is biased for Class A operation so that collector current will

Single-Sideband Transmitter Blocks

flow throughout the cycle, maintaining the sinusoidal form of the signal from the mixer.

Automatic Limiting Circuit

Waveshapes of an SSB signal can be observed on an oscilloscope using the setup and equipment used when making AM measurements. Here, however, two audio tones of equal amplitude must be fed simultaneously to the microphone input to generate a useful display at the output. The display on an oscilloscope will appear as in Fig. 5-26. Determine the

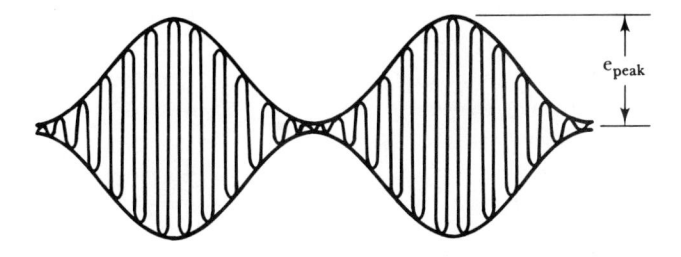

FIG. 5-26 Envelope due to two-tone audio signal at MIC input of SSB transceiver.

voltage shown in the figure from the display. Peak power is related to the peak voltage by the equation

$$\text{PEP} = (0.707\, e_{peak})^2/52 = e_{peak}^2/104$$

if a 52-Ω load is used at the output. The peak power must be limited to 12 W, so the maximum peak voltage that should be available for display on the oscilloscope is

$$e_{peak} = 10.2\, (\text{PEP})^{1/2}$$

Special automatic limiting circuits (ALC) are designed for SSB applications to keep the voltage, and hence power, within acceptable limits. One circuit of this type is shown in Fig. 5-27.

RF voltage at the antenna jack is rectified. Pulsating direct current at the output of diodes D1 and D2 is filtered by capacitor C. The size of this dc voltage depends upon the amount of RF voltage at the antenna. The amount of current flowing through the collector circuit of Q1 is related to the size of this voltage. Collector current and RF voltage at the antenna jack increase simultaneously. As the collector current rises, the voltage across load resistor R2 also increases. Consequently, voltage at the collector drops. We then have the situation where an increase in signal at the antenna jack produces a corresponding decrease of voltage

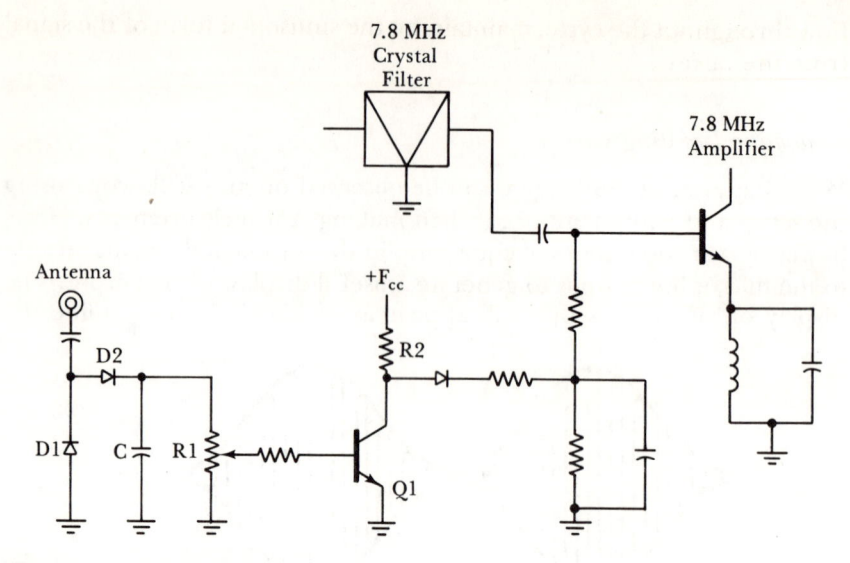

FIG. 5-27 ALC circuit.

at the collector of Q1. This voltage is applied to the bias circuit at the base of the amplifier transistor located just after the 7.8-MHz crystal filter in the transceiver.

Should the RF at the antenna jack increase beyond a predetermined value, the dc bias voltage at the base of the 7.8-MHz amplifier becomes less positive. The gain of the amplifier is reduced. Consequently, the amplitude of the 7.8-MHz signal at the output of this amplifier is reduced. Being a determining factor in the size of the output signal, the reduced output from the 7.8-MHz amplifier triggers a reduction of the voltage at the antenna jack. PEP is restored to maximum acceptable levels.

The ALC control is set so as to limit PEP to within the dictates of the FCC. For a PEP of 12 W, e_{peak} should be a maximum of 35.4 V. Adjust control R1 so that the output voltage at the antenna cannot exceed this under any signal conditions.

6

Troubleshooting the Receiver

Conventional AM broadcast band receivers are nothing new to the service technician. He has handled them ever since he learned which end of a soldering iron never to touch. The receiver section of a transceiver is very similar to that of an ordinary radio. They differ primarily in the frequencies accommodated. In addition, there are a number of embellishments peculiar to CB radios. These have been described thoroughly in Chapter 2, and should be reviewed carefully before proceeding with the discussion to follow. Note especially the blocks in the single- and dual-conversion receivers in Figs. 2-5 and 2-7, respectively.

There is major disagreement among CB enthusiasts as to whether the dual- or single-conversion receiver is the most desirable. Dual-conversion circuitry is used where image frequency rejection is most important. For example, when the channel selector switch is set to channel 2, the carrier frequency is 26.975 MHz. If the IF used in the radio is 0.455 MHz (455 kHz), the local oscillator must supply a signal at 26.975

+ 0.455 MHz, or 27.430 MHz. Now, 27.885 MHz is also 0.455 MHz away from the 27.430-MHz local oscillator frequency. If there is transmission at 27.885 MHz, the RF here beats with the 27.430 MHz from the oscillator to also produce 0.455 MHz for the IF amplifier. Signal broadcast at the 27.885-MHz image frequency would consequently interfere with reception of signal from channel 2. One or more tuned RF amplifiers are usually in the circuit preceding the mixer to minimize this interference. But it is difficult to eliminate image frequencies that are only 910 kHz (2 × 455 kHz) away from the desired signals. Image frequencies can be filtered easily in the RF stages, if their frequencies are far removed from those of the CB band. This requirement can be fulfilled by using a high IF. The desired and image frequencies would then be sufficiently removed from each other (twice the intermediate frequency) to facilitate suppression of the interfering signals.

Dual-conversion systems use high IF filters immediately after the RF and HF mixer stages. Here the CB and image frequencies are far apart. Because the image frequency is easily suppressed, only the CB signal passes through the RF stages to the IF filter. The high IF is then converted to a low IF in a second mixer stage. Here signal from the high IF beats against signals from the second local oscillator to produce the low IF.

Advocates of single-conversion systems make two valid claims. First, the CB band is less than twice the IF bandwidth (less than 910 kHz wide), so there is no transmission on that band at an image frequency. Second, when noise blanking circuits are used, the circuit that blanks the noise pulse must be activated at the instant the noise pulse is present. This coincident condition is difficult to attain when IFs are low. You can, however, have the best of both worlds by using high IFs in single-conversion circuits. Which system is best? If you get a good transceiver, it probably does not matter!

The receiver section is an integral part of a CB radio. It cannot be completely separated from the transmitter. Much of the circuit common to both sections has already been discussed in Chapter 5. Here we shall use the views of the receiver section shown in Figs. 2-5 and 2-7, and supply the details for the boxes.

Radio-Frequency Amplifier

Received signal from the antenna is magnified in the RF amplifier before it is applied to the mixer. Gain can be increased substantially if more than one RF stage is used. The added gain is seldom required. A drawback of excessive gain is that extremely large signals can overdrive succeeding stages including the mixer, and produce spurious signals in the receiver.

Radio-Frequency Amplifier

Tuned circuits are usually placed at the input and output of the RF amplifier. These tuned circuits substantially narrow the band of frequencies passing through this amplifier to subsequent stages. Interference from image frequencies is thereby minimized. Various versions of RF amplifiers used in CB radios are shown in Fig. 6-1.

One or two diodes are shown at the inputs of each RF amplifier circuit. As discussed in previous chapters, voltage across a forward-biased diode is limited to about 0.2 V for germanium and about 0.7 V for silicon devices. As applied to the circuits in Fig. 6-1, the diode limits voltage at the input of the RF transistor to the forward-biased voltage of the diode used. The input junction of the transistor is consequently protected from destruction by large signal and noise pulses that may be on the antenna.

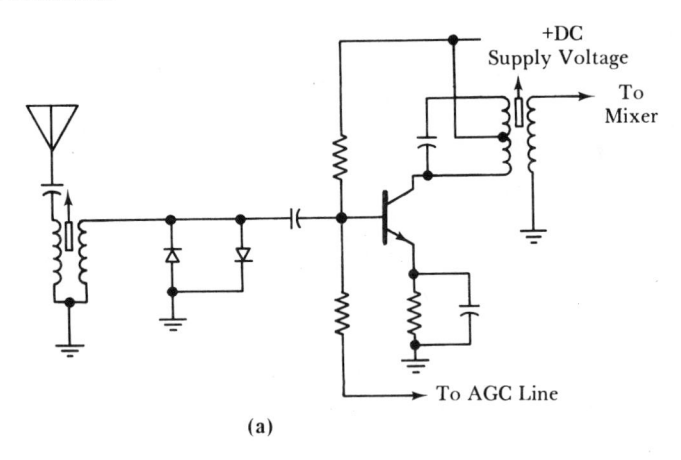

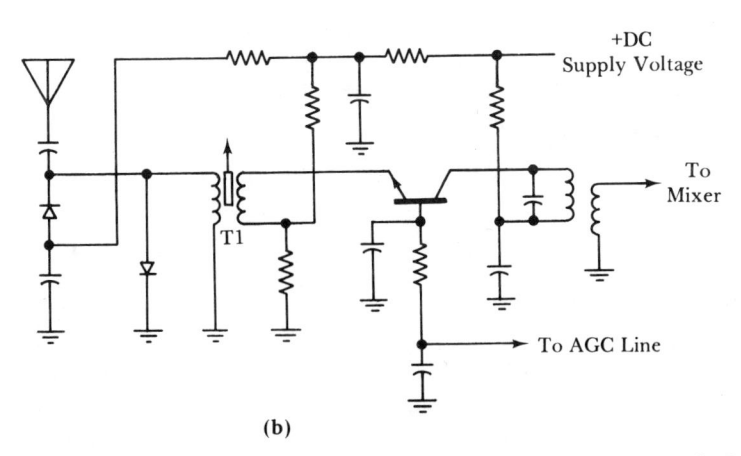

FIG. 6-1 RF amplifier. (a) Common emitter circuit using bipolar transistor, (b) Common base circuit using bipolar transistor.

(c)

(d)

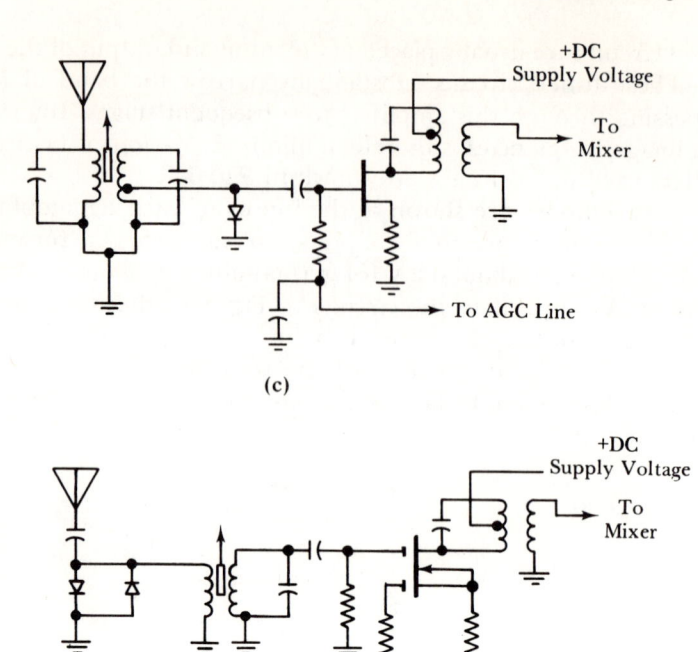

FIG. 6-1 (*Continued*) (c) Common source circuit using junction FET, (d) Common source circuit using insulated gate FET.

Noise pulses are very fast. Diodes may react too slowly to these pulses and thus be incapable of protecting the transistor. At times, noise pulses pass through the circuit too quickly to turn on the diode. Pulses are limited most effectively through use of the circuit in Fig. 6-1b, for here the diodes are forward biased by the supply voltage and are thus always in an "on" state.

Two diodes are used in Figs. 6-1a, b, and d, limiting both positive and negative noise pulses. Only one diode is required for the FET circuit in Fig. 6-1c. Here only positive pulses are limited by the diode. It is assumed that the negative pulses cannot destroy the JFET gate junction.

RF amplifiers are functional elements in the AGC chain. AGC voltage is applied to the base or gate of the transistor. It affects the gain of the stage in such a manner as to keep its output constant regardless of the strength of the received signal at the antenna. The circuit in Fig. 6-1d is most interesting because a dual gate IGFET is used. Here AGC is applied to one of the gates while the modulated RF is applied to the second gate.

Radio-Frequency Amplifier

All tuned circuits should be aligned for proper reception of the entire 27-MHz CB band. Considering the 23-channel transceivers, a signal at the center of the CB band is at 27.11 MHz. All tuned circuits should be peaked at 27.11 MHz, and checks should be made to confirm that all frequencies from 26.962 to 27.258 MHz can pass through this stage with gain equal to that of the 27.11-MHz signal. This is to reassure you that all sidebands as well as the carriers of channels 1 through 23 are reproduced properly. On 40-channel transceivers, the center of the band is at 27.185 MHz, and the tuned circuits should pass frequencies from 26.962 to 27.408 MHz without attentuation.

Troubleshooting procedures to be applied here differ little from those employed when checking any stage of gain. First, measure the voltages. If gain is low, the dc voltage used to power the stage may be insufficient. Should this be the case here, check the resistors and capacitors composing the decoupling circuits of the B+ line, as described in Chapter 5. If voltages are proper, a bypass capacitor in one of the emitter or source circuits, or in the bias circuit of the RF transistor, may be at fault. Check these by shunting those in the circuit with another one of like value. Gain increases immediately when an open bypass capacitor is shunted by a good one.

Transistors may turn defective in RF and IF stages. Due to the low resistance of the coils and transformers in the circuit, transistors must be removed from the printed circuit board before they can be tested. Do this cautiously and carefully, taking heed not to apply excessive heat to the leads. Once removed from the circuit, the transistor can be checked using one of the testers described in Chapter 4.

Diodes can also fail. Check them in a transistor tester or use an ohmmeter. Resistance across the device should be relatively high when the cathode is connected to the more positive lead of the meter and low when the anode is connected to that lead.

RF transformers usually have ferrite slugs as tuning devices. They are adjusted by turning them into or out of the coil form. Slugs can move in the form and upset the adjustment, or they can break by being jarred or manhandled. You can use a dip meter to determine if the transformer must be realigned. When using a dip meter for these tests, be sure to first remove all power from the transceiver. Place the meter near the RF transformer in question and note if it is resonating at the specified frequency. The resonant frequency should vary as the slug is readjusted. Should the frequency remain unchanged, you have either a cracked slug or another defect in the transformer. In some cases, you can rectify the situation by simply replacing the slug. If you do this rather than replacing the entire transformer or coil, be sure that the replacement slug is identical in size and material to the original. If an exact duplicate of the original slug is not available, the entire transformer must be replaced.

Some manufacturers add an RF level control as a refinement to the stage. The function of this appendage was discussed in Chapter 2. If you consider the circuits in Figs. 6-1a, c, and d, the level control is merely a potentiometer in the emitter or source circuit of the RF transistor. Should the control be set so that its maximum resistance is in the circuit, the gain of the RF amplifier is at a minimum owing to the large amount of emitter feedback. When presenting 0 Ω to the emitter or source circuit, feedback is minimized so that the RF gain is at a maximum.

A two-position switch rather than a variable control is used on some transceivers. The two settings of the switch are LOCAL and DISTANCE. Higher resistance (for reduced gain) is obviously introduced in the emitter or source circuit when the switch is set to LOCAL than when it is set to DISTANCE. Although it can do its job adequately when wired in the RF stage, some manufacturers place the switch in the emitter or source circuit of the mixer.

Switches are mechanical devices and are prone to failure. If gain is excessively low in either or both switch positions, suspect the switch. Performance can usually be restored to normal by spraying the contacts with a cleaner.

Local Oscillator, Mixer, and Intermediate-Frequency Stages

Amplified signal from the RF stage is combined in the mixer with signal from the local oscillator. The difference frequency of the two signals is chosen as the intermediate frequency. It is amplified by one or more IF stages. In dual-conversion receivers, the initially selected IF is fed through a second mixer. Here it is combined with the signal from a second local oscillator to produce a lower IF.

Oscillators and mixers were discussed in Chapter 5, to which we refer you for all circuit and troubleshooting details.

Frequency Synthesizers

A frequency synthesizer used in 23-channel single-conversion receivers was described with reference to Fig. 5-2. Local oscillator frequencies for the receiver circuit are derived here from the combination of signals from the receive and HF oscillators. They are mixed in a synthesizer. The resultant frequencies are combined in the receiver mixer (or first detector of Fig. 2-5) with received RF signals to form the 455-kHz IF.

As for dual-conversion receivers, signals from the HF oscillator are combined in the HF mixer of the receiver (see Fig. 2-7) with received RF signals to produce the high IF. Six different frequencies can be produced by the HF oscillator in 23-channel transceivers. In the Royce Model 1-602A transceiver shown in Fig. 5-5, the frequencies produced by the oscillator are 37.60, 37.65, 37.70, 37.75, 37.80, and 37.85 MHz. The 37.60 MHz beats against the received RF signals from channels 1 through 4. The 37.65 MHz beats against the received RF signals from channels 5 through 8, and so on down the line. By combining each of the six frequencies from the oscillator with signals from up to four of the received channels, four different IF frequencies can be formed. They range from 9.595 to 10.635 MHz. The HF IF filter must be capable of passing this band of frequencies. The center of the band is 10.615 MHz. IF filters must be capable of passing a 40,000-Hz-wide band, or 20,000 Hz on both sides of 10.615 MHz, without any real attenuation at the extreme ends of the band. On each channel, the particular frequency passing through the first IF stage must then beat in the LF mixers with one of the four frequencies available from the receive oscillator to produce the 455 kHz selected and amplified by the low IF amplifier.

Intermediate-Frequency Stages

IF stages are usually lower-frequency versions of the RF stages (less the protecting diodes) discussed above. They may use L–C circuits, mechanical filters, or both as the frequency-determining components. The various circuits should be adjusted to resonate at the frequency prescribed by the manufacturer of the equipment. For single-conversion sets and for the IFs after the second mixer in dual-conversion radios, this frequency is usually 455 kHz. Troubleshooting procedures are similar to those prescribed above for the RF stage.

Delta Tune

Delta tune as a feature was described in Chapter 2. It is used to compensate for any deviations of the received signal from its assigned carrier frequency. To compensate for this discrepancy, a local oscillator frequency in the receiver is altered so that, when its signal mixes with the straying received signal, the exact required IF is produced.

In dual-conversion receivers, delta tune is accomplished by changing the frequency of the second local oscillator.

Three versions of switched delta tune circuits are shown in Fig. 6-2. In each circuit, the oscillator frequency required when reproducing an accurately received signal is selected when the switch is set to position 2. Oscillator frequencies 1500 Hz above and below the one at dead center are selected in the alternative settings of the switch. In the circuit of Fig.

FIG. 6-2 Delta tune circuits.

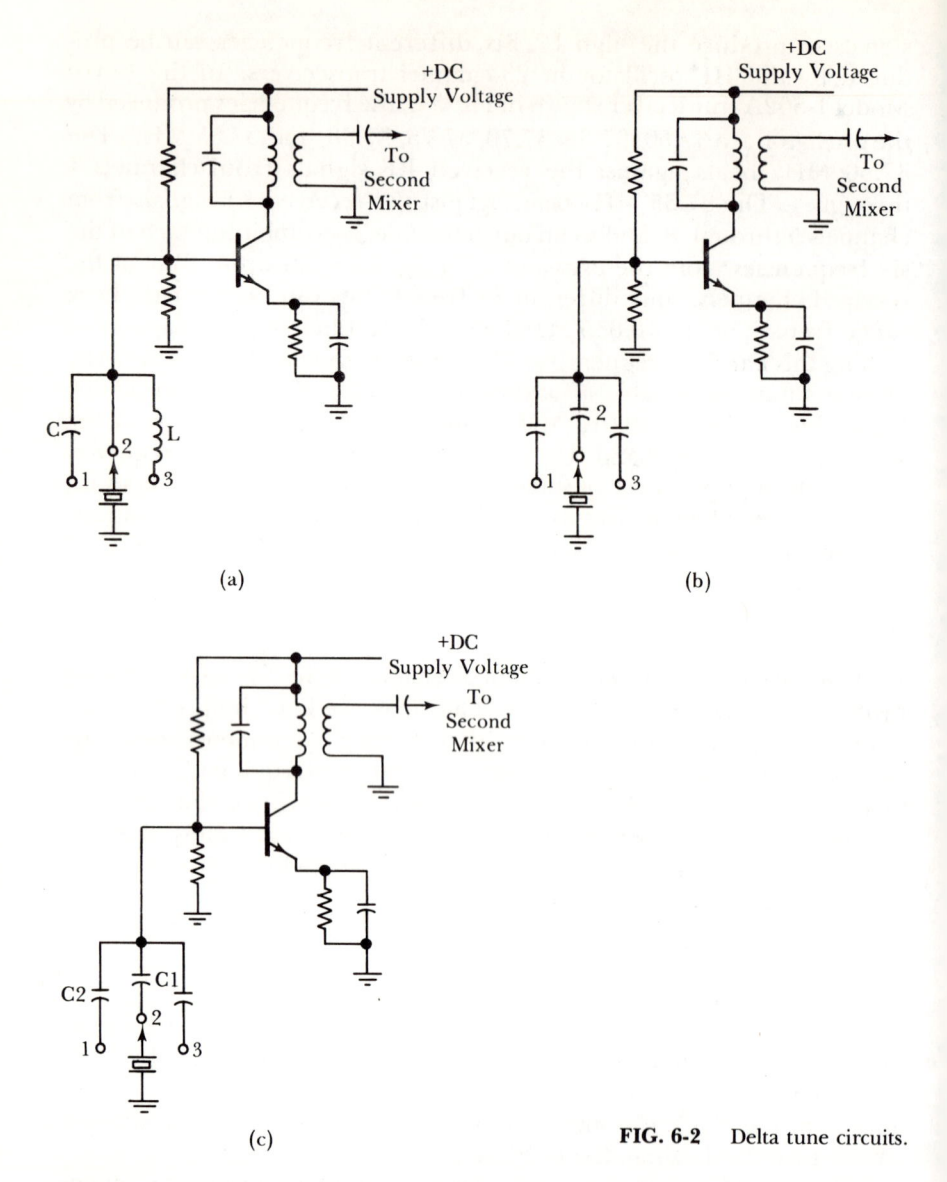

6-2a, a capacitor is placed in series with the crystal to alter the center frequency in position 1 of the switch. An inductor performs this function in position 3. Only capacitors are used in the circuits in Figs. 6-2b and c. A practical circuit utilizing the delta tune feature is shown in Fig. 5-5. This circuit uses the arrangement in Fig. 6-2a.

Rather than using a switch to select one capacitor in each position, as shown in Fig. 6-2b, a variable component can be wired in series with the crystal. Its reactance can be varied by a knob at the front panel. Frequency can be changed continuously as the reactance is varied, rather than in three steps. In SSB units, this continuous frequency variation capability is essential, as the adjustment is critical. Here the control is set for the cleanest and most intelligible reception. Consequently, the control is referred to as a *clarifier*. SSB receivers will be detailed later in this chapter.

Detectors

Output signals from the IF stages look very much like the modulated RF shown in Fig. 2-8c. Here, however, the "carrier" is not in the 27-MHz range, but is at the intermediate frequency.

Because there is audio on both the top and bottom halves of the modulated RF (or IF) signals, the two audio components cancel. In order to get audio intelligence from the AM pattern, either the top or bottom half of the signal must be eliminated. By passing the modulated IF through a diode, all IF signals above or below (depending upon the direction of the diode in the circuit) the horizontal axis are eliminated. Only the modulated upper (or lower) portion of the IF passes through the diode. Audio can be retrieved from the modulated IF signal by simply bypassing all IF so that only audio remains.

A circuit used in practically all CB radios is shown in Fig. 6-3. Modulated IF signals pass through diode detector D1. The IF carrier is

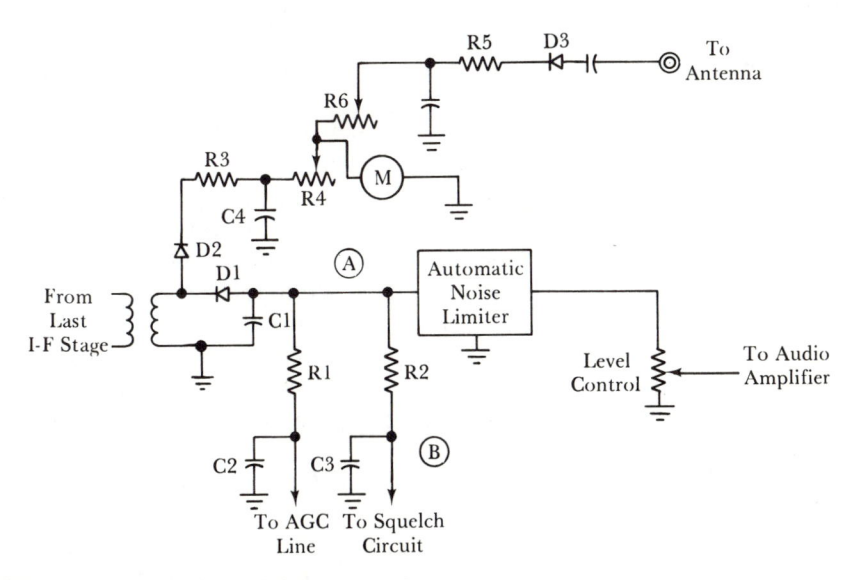

FIG. 6-3 Commonly used defector circuit.

bypassed by C1, and only audio remains at point A in the circuit. Audio passes through the automatic noise limiter (ANL) to the level control, and from there to the audio amplifier for reproduction through the loudspeaker.

There are several other circuits activated by the audio across C1. It is fed through R1 and converted into a slowly varying dc voltage through the action of R1 and C2. This dc voltage is applied to the bias circuits of the RF and IF amplifier to control their gain. The varying dc voltage is at a proper level to equalize the gain of the overall transceiver for all input signal levels. Much more dc voltage is fed back on this line, the AVC line, when strong signals are received than when the amplitude of these signals is low. High dc voltages limit the gain of the IF and RF stages while the low voltages do not. Owing to this AVC feedback action, output signals are of more or less equal amplitudes irrespective of the strength of the signal at the antenna and RF stage.

Audio at point A also passes through resistor R2. It is filtered by the R2–C3 network so that the dc voltage across C3 varies slowly with the audio voltages. This slowly varying dc voltage, a voltage changing with the amplitude of the received RF, is applied to the squelch circuit for gating it on and off.

Signal at point A passes through a noise-limiter circuit before being amplified. Here fast noise pulses, such as those from the ignition system of the automobile, are not permitted to pass through to the level control and to the balance of the audio amplifier. Actually, the noise limiter blocks the passage of all signals from point A to the level control for the instant that noise pulses are present. Because the time of the pulse is very short, the "moment of silence" is not noticeable as an absence of audio intelligence.

Returning to the modulated output from the IF, note the circuit stemming from diode D2. Here the signal is detected by the diode, filtered by the R3–C4 combination, and passed through potentiometer R4 to meter M. In the receive mode, it serves as an S-meter.

Note that a similar circuit from the antenna jack is completed through diode D3, R5, and potentiometer R6 to the meter. This circuit has significance only when transmitting. Now the deflection of the pointer on the meter indicates the power that is being fed from the transmitter circuit to the antenna jack. The potentiometers are used to calibrate the meter in the two modes of operation.

Automatic Noise Limiter

A schematic of a frequently used ANL circuit is shown in Fig. 6-4. Point A and the level control are also shown in Fig. 6-3. The cathode end of diode D4 is close to ground potential because of the circuit through

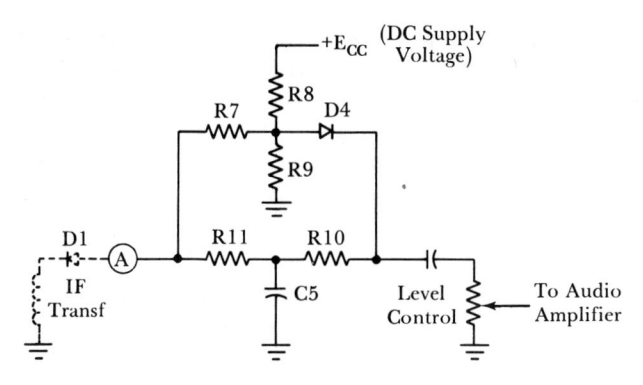

FIG. 6-4 Noise limiter.

resistor R10, R11, diode D1, and the IF transformer. Positive voltage with respect to the cathode of D4 is applied to its anode from the $+E_{CC}$ supply through the action of voltage divider resistors R8 and R9. D4 is almost always in the "on" state because of the polarity of the voltage. Signal passes readily from point A through R7 and D4 to the audio amplifier.

In the meantime, capacitor C5 is charged through R11 by the audio at point A. Because of the orientation of D1, voltage across the capacitor is negative with respect to ground. Voltage across C5 varies slowly as speech signal levels fluctuate. But an instantaneous high amplitude noise pulse is too fast to produce any voltage changes across C5.

Negative voltage at the junction of C5 and R10 is relatively low. The cathode of D4 is maintained at this low dc voltage at C5 through resistor R10. A large but short-lived negative noise pulse at point A never reaches the cathode of D4 because of the time constant of R11 and C5. But it does reach the anode of D4, making it more negative that the cathode. The diode is turned off by the instantaneous pulse. The path of the audio from point A to the potentiometer through the diode, is opened for the instant that the noise is present.

ANL circuits can be easily switched in or out of the signal path. All that the ANL on–off switch usually does is short diode D4 to defeat the limiting action. By opening and closing this switch, you can determine just how effective the circuit is in eliminating the various types of noise pulses that enter the receiver.

Noise Blanking

The ANL is at the output of the detector stage, but noise can also be prevented from passing through earlier stages of the receiver. One scheme uses a noise blanking circuit. It differs from the ANL by sensing all types of noise at the antenna rather than at the second detector. Here the noise is amplified, detected, and applied to special diodes in the IF chain. Direct current due to this noise is sufficient to turn the diodes off for the time that the noise is present.

A typical noise blanker or silencer circuit is shown in Fig. 6-5. Diodes D1 and D2 are turned on by voltage from power supply $+E_{CC}$ through the viltage divider formed by R1 and R2. In the circuit shown below this, noise from the antenna is amplified by a noise amplifier and detected by diode D3. When noise is present, positive pulses are available at the cathode of D3. The pulses are applied to the center tap of T2. Diodes D1 and D2 are cut off by these pulses. At the time that the noise is present, signal is thereby prevented from passing through the mixer to the IF amplifiers and is consequently restricted from the balance of the receiver.

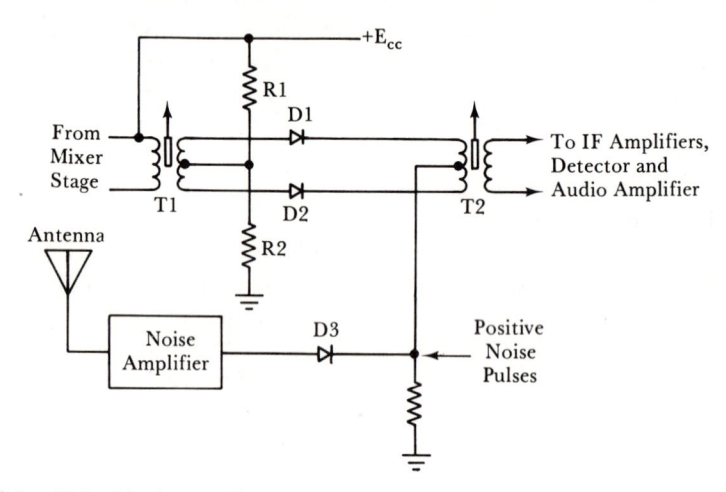

FIG. 6-5 Noise blanker or silencer.

Troubleshooting Detectors and Noise-Deleting Circuits

The biggest troublemaker in any of these circuits is the diode. Should any of these circuits fail to operate, carefully disconnect one lead of the suspected diode and check the forward to reverse resistance ratio with an ohmmeter. Resistance ratios of RF diodes can be as low as 10:1; those of audio signal devices should be many times higher.

Precautions should be taken not to overheat the diode when removing it from the circuit. It is best to hold the lead involved with long-nose

Squelch Circuit 141

pliers while applying heat to the pad on the printed circuit board. Since the pliers are between the body of the diode and the source of heat, it acts as a sink by drawing heat to itself rather than letting the heat reach the body of the delicate device.

Squelch Circuit

Returning to Fig. 6-3, it was indicated earlier that filtered dc voltage at the junction of R2 and C3 is used to activate the squelch circuit. The dc voltage at this juction is amplified by a squelch gating transistor. Its output is connected to one of the transistors in the audio amplifier chain, where it is applied in such a manner as to limit audio gain when the received signal is weak.

A circuit designed to perform this function is shown in Fig. 6-6. Audio from the level control in Fig. 6-3 is fed to the base of audio amplifier transistor Q1. Q1 does not amplify or conduct current unless there is sufficient dc bias voltage at its base to turn it on. The size of the dc voltage at the base of Q1 depends upon the amount of voltage at the collector of the squelch gating transistor Q2. This, in turn, is related to the dc voltage at its base.

The base of Q2 has two sources of dc voltage. One is available from the wiper of the squelch control, for it is wired across the $+E_{CC}$ supply; dc voltage for the base of Q2 is also derived from the filtered audio at point B of Fig. 6-3.

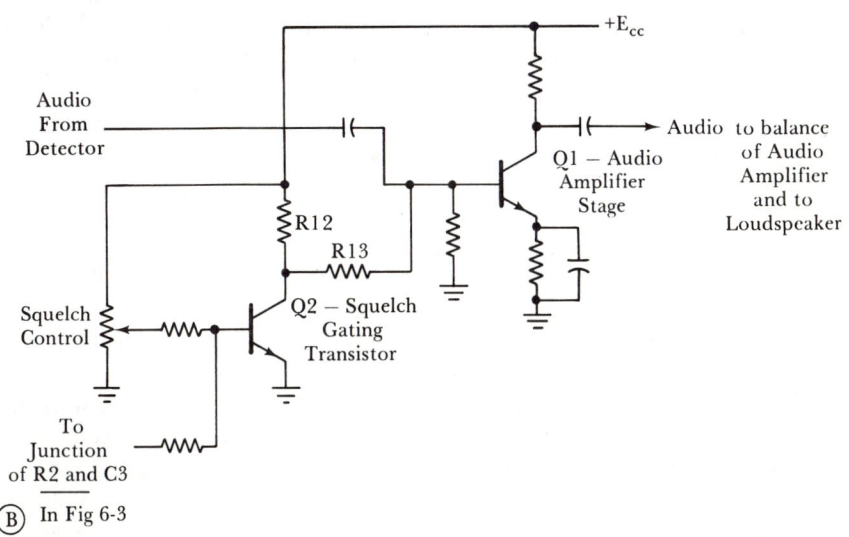

FIG. 6-6 Squelch gating circuit.

For the moment, assume that there is no voltage at point B, indicating that no signal is being received. Bias on Q2 depends solely on the voltage selected for its base by the setting of the squelch control. When the wiper is near the $+E_{CC}$ end of the control, a large amount of current flows through the base and collector circuits of Q2 and through the load resistor in its collector, R12. Voltage across the resistor is large, for it is equal to the collector current multiplied by the value of the resistor. Subtracting this big voltage from $+E_{CC}$ leaves very little voltage at the collector of Q2. When applied through R13 to the base of Q1, this voltage is too low to turn on the transistor. As it is off, no audio can pass through Q1 to the balance of the audio amplifier circuit. There is no audio through the loudspeaker.

Next assume that the wiper on the control is set at the ground end of the control so that no dc bias voltage is at the base of Q2. Q2 is turned off. Only a minute current flows through its collector and through R12. As there is practically no voltage developed across R12, just about the entire $+E_{CC}$ supply voltage is at the collector of Q2. It is applied to the base of Q1 through R13, turning on that audio amplifier stage. All weak and strong audio signal can now pass through this transistor to the balance of the audio circuit.

The wiper on the squelch control is normally set to the point where Q1 is turned off when the received signal is weak and noise is excessive.

Assume for the balance of this discussion that the wiper on the squelch control is set to the point where it does not allow weak signals to pass through Q1. Q2 is turned on. When a reasonably strong signal is received, there is a negative dc voltage at point B. Should the amplitude of the signal be sufficient, enough negative direct current is available at point B to reduce the bias voltage at the base of Q2. Consequently, the collector current in this transistor is either diminished or entirely cut off. Voltage at the collector of Q2 rises, turning on transistor Q1. Signals can then pass through this transistor. Noting the sequence, Q1 is cut off (or the output signal is squelched) until the received signal is sufficient to be intelligible. There is no sound through the loudspeaker unless there is sufficient signal for proper reception.

The circuit in Fig. 6-6 is essentially the one used in most CB radios. To be sure, there are many variations. For example, the amplitude sensing input voltage may be taken from the emitter circuit of any RF, mixer, or IF stage instead of from point B in Fig. 6-3, and fed to the base of Q2 in Fig. 6-6. The transistor selected as the source of the amplitude sensing signal is any one of the RF amplifiers used as a return for the AGC line, because its emitter voltage decreases while RF is being received.

If the squelch circuit is to work properly, the amount of dc voltage at the output of the gate circuit must change radically with the level of

Squelch Circuit

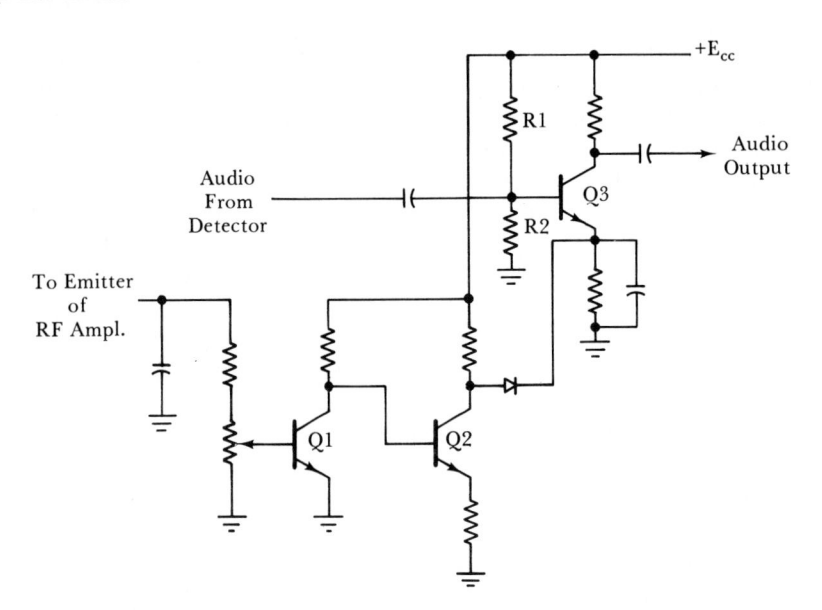

FIG. 6-7 Squelch gating circuit with gate voltage applied to emitter of audio stage, Q3, from squelch amplifier Q2.

the signal received. This is best accomplished by incorporating a stage of dc gain at the output of the squelch gate transistor (Q2 in Fig. 6-6). You will find this extra stage on many CB radios.

The base circuit of the first audio stage is not the only one that can be used for turning the device on and off. As shown in Fig. 6-7, squelch gate voltage can be applied to the emitter to achieve the same goal. Here diode D is an important component of the functional circuit. It is placed between the collector of gating transistor amplifier Q2 and the emitter of audio amplifier stage Q3. The anode of the diode is at the collector of Q2. When voltage at its collector is low, the diode is reverse biased. Under these conditions, the emitter voltage of the audio stage is not affected by voltage at the collector of Q2. Should the voltage at its collector rise, the diode is turned on. High voltage at the collector of Q2 is applied to the emitter of the audio amplifier through diode D. Q3 is cut off by this voltage.

When signal is being received, low voltage available at the emitter of the RF amplifier is applied to the base of Q1, reverse biasing that transistor. Q1 is turned off. The consequent high voltage at the collector of Q1 and base of Q2 turns on Q2. High collector current through that transistor forces its collector voltage down. Diode D is turned off because the voltage at the collector of Q2 is less than that at the emitter of Q3. Q3 is forward biased through R1 and R2. Audio passes freely through Q3.

Should the signal input to the receiver be weak and noisy, or even if it is entirely absent, voltage is high at the emitter of the RF amplifier. When applied to the base of Q1, the voltage turns on this transistor. Going through the steps outlined in the previous paragraph, you will determine that the voltage at the collector of Q2 is high under these conditions. Diode D is turned on, applying the positive collector voltage at Q2 to the emitter of Q3. Q3 is reverse biased. None of the noise or weak signals in the RF and IF stages can pass through Q3 and the balance of the audio amplifier.

Troubleshooting procedures for any squelch circuit start with voltage measurements at the various terminals of the transistors and diodes. Voltages should be measured with and without signal reception. To simulate a no-signal condition, connect a dummy load at the antenna terminal. Very little signal will then pass from the antenna jack to the receiver section. If you want a strong input signal for testing purposes, couple the output from an RF generator to the antenna circuit. Under both conditions, voltages at the transistor and diode terminals should be as described. If there are discrepancies, check the semiconductors using one of the in-circuit testers described in Chapter 4. Besides the transistors, capacitors should also be checked. Test them for opens or shorts, especially if voltages remain constant regardless of the strength of the signal at the antenna jack.

Audio Amplifier Circuits

Modulators were discussed thoroughly in Chapter 5. The same audio amplifier used for modulating the RF in the transmit mode of operation is used here for reproducing signals from the detector. Signal from Q3 in Fig. 6-7 is usually fed to the balance of the audio amplifier section for

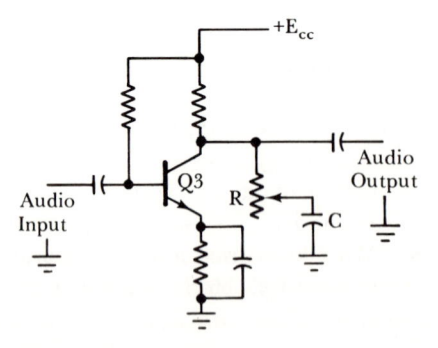

FIG. 6-8 Audio amplifier with tone control circuit. Many different arrangements are possible.

reproduction through the loudspeaker. As the circuit was detailed in the previous discussion, along with troubleshooting procedures, no further elaboration is required here.

Tone controls are appendages on some CB radios. These are capacitor connected to a potentiometer in the collector circuit of Q3 (see Fig. 6-7), as shown in Fig. 6-8. Treble is attenuated as the wiper on potentiometer R approaches the top of the resistance element. Only then is the capacitor fully in the circuit and capable of limiting the upper end of the audio band.

Aligning the Receiver

Before beginning to describe an actual alignment procedure, it is imperative to realize that we can only generalize. If you have specific alignment instructions from the manufacturer of the transceiver, follow them rather than the methods in this section. Also note that the procedure detailed here is similar to the one used for aligning an ordinary AM broadcast band radio. If you have done this before, the following is simple indeed. Finally, it is assumed here that the transmitter section has been properly aligned using either the procedure in Chapter 5 or that provided by the manufacturer of the equipment. It is taken for granted that all oscillators are "perking along" at the proper frequencies.

Using the block diagram in Fig. 2-7 as a reference, the following is a method for aligning a dual-conversion receiver:

1. Connect a 50-Ω dummy load to the antenna jack.
2. Disable the LF local oscillator by either removing the crystal or by disconnecting the dc voltage source from the oscillator transistor.
3. Monitoring the frequency of an RF generator with a digital counter, set the generator to supply the low IF frequency, usually 455 kHz. Feed the output from the instrument through a 1000-pF capacitor to the input of the LF mixer. Connect the shielded lead from the cable to a ground on the transceiver.
4. Connect a TVOM to the AGC line at the detector. Set the meter to a dc voltage range.
5. Peak all IF transformers. Start at the one closest to the detector. Work your way back to the transformer at the output of the LF/mixer. When peaked, the TVOM indicates a maximum dc voltage. If the meter reading does not vary as you

adjust the IFs, the output voltage from the generator is too high. Reduce it and continue adjusting the IF transformers for a maximum reading on the TVOM.

6. Reset the frequency of your generator first to provide a signal 5 kHz below the center IF frequency, and then to provide a signal 5 kHz above it. The TVOM will read about the same voltage at both settings if the response of the IF system is symmetrical around the center frequency. It may be necessary to increase the output from the generator in order to get a reading on the TVOM. If both output voltages are not the same, retune the stages slightly until symmetry within about 20 percent is achieved.

7. Put the LF local oscillator back into operation and disable the HF local oscillator using one of the methods described in step 2.

8. Disconnect the "hot" lead of the signal generator from the input to the LF mixer and connect it to the input of the HF mixer. Set the generator to the frequency specified at the output of the HF mixer, or to the frequency of the HF IF amplifier, if there is one in the transceiver.

9. Peak the transformer(s) located between the two mixers in the circuit, noting the output on the TVOM. On 23-channel transceivers, check voltages on the TVOM at frequencies 20,000 Hz above and below the center frequency. If there is a substantial difference, readjust the alignment so that voltages at all frequencies discussed are at about the same level. Now check voltages at 21,500 Hz above and below the center frequency. Both voltages should be about equal. If they are not, additional adjustments of the tuning are required until this requirement is satisfied. Required bandwidths differ on the various types of 40-channel transceivers. Manuals of specific 40-channel CB radios should be consulted for the applicable information in each case.

10. Restore the HF local oscillator to normal operation.

11. Disconnect all leads from the generator to the transceiver. When aligning a 23-channel radio, set the transceiver to channel 12 and adjust the generator so that it provides exactly 27.105 MHz, as noted on the monitoring digital counter. Use channel 19 and 27.185 MHz for 40-channel transceivers. Drape leads from the generator around the dummy load at the antenna jack.

Single-Sideband Receivers 147

12. Peak all coils in the RF amplifier stages.

13. Reset your transceiver first to channel 1 and then to channel 23, if you are working on a 23-channel radio. Set the generator to 26.965 MHz when the channel selector switch on the transceiver is set to channel 1 and to 27.255 MHz when it is set to channel 23. The TVOM will indicate approximately the same voltage when the transceiver is set to both channels 1 and 23, if the response of the RF-tuned circuits is symmetrical. If the readings are not the same, retune the RF circuits slightly until symmetry within about 10 percent is achieved. In fact, the readings on the TVOM should be within 10 percent of each other on all three channels used in this procedure. On 40-channel radios, use the frequency of channel 1 at 26.965 MHz at one setting and the frequency of channel 40 at 27.405 MHz in the second setting when performing the above measurements.

The procedure just described was devised for aligning dual-conversion-type transceivers. If you are working on a single-conversion radio, omit steps referring to the HF mixer. Skip steps 7 through 9, and in step 10, change "HF" to "LF."

Single-Sideband Receivers

A dual-conversion receiver section of a transceiver is shown in Fig. 2-7. In its essentials, this block diagram can also be used to represent the SSB receiver circuit. It is elaborated upon in Fig. 6-9 to indicate how the various high- and low-frequency local oscillator signals are applied, as well as to denote other peculiarities in the SSB mode of operation.

Received signal is magnified in the RF amplifier stage and fed to the HF mixer. Tracing the RF signal for channel 1, as was done in Chapter 5, the center frequency of the LSB band of signals is at 26.9635 MHz, and the center frequency of the USB band of signals is at 26.9665 MHz. These received signals are combined with the local oscillator frequencies formed in the frequency synthesizer.

As was the case for the transmitter, 11.705 MHz is supplied by the six-frequency oscillator when channel 1 is selected. This is combined in the frequency synthesizer or mixer with 7.4585 MHz from one four-frequency oscillator when the LSB is selected and with 7.4615 MHz from the second four-frequency oscillator when the USB is selected. The output frequency chosen from the synthesizer is the sum of 11.705 MHz and either 7.4585 MHz or 7.4615 MHz, depending upon whether the

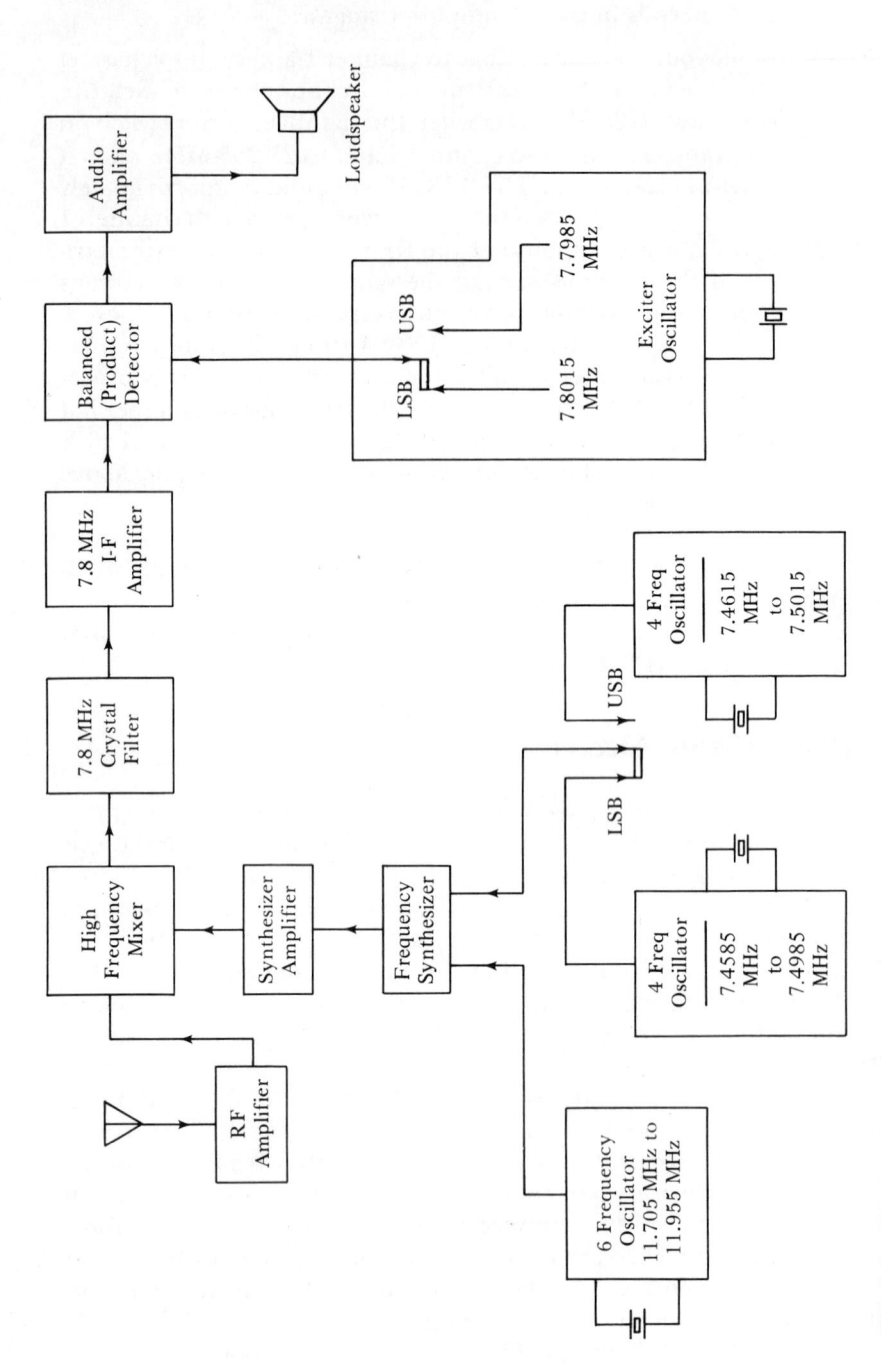

FIG. 6-9 Block diagram of SSB receiver.

LSB or USB is being used. Thus the output from the frequency synthesizer (or oscillator mixer) and its amplifier is 19.1635 MHz when the mode switch on the front panel is set to LSB and 19.1665 MHz when the switch is set to USB.

Output from the synthesizer is combined with the received RF in the HF mixer. On channel 1, this is 26.9635 − 19.1635 = 7.8 MHz when using the LSB, and 26.9665 − 19.1665 = 7.8 MHz when using the USB. Regardless of the channel used, the proper oscillator frequencies are generated and combined in the synthesizer with the received signal to form 7.8 MHz at the output from the HF mixer.

As was the case with the transmitter signal, IF is exactly at 7.8 MHz when the received signal has 1500-Hz audio information. There is actually a narrow band extending to about 1500 Hz on either side of the 7.8 MHz that must be passed to accommodate all transmitted audio frequencies from about 300 to 3000 Hz. This band is selected by the crystal filter, amplified by the IF stages, and fed to a product or heterodyne detector. Here a frequency from an exciter oscillator is combined with the narrow IF band around the 7.8 MHz. IF and exciter oscillator frequencies are filtered from the output. The remaining audio is fed to the ensuing stages to be amplified and applied to the loudspeaker.

For use on the LSB, the exciter oscillator frequency is at 7.8015 MHz; it is 7.7985 MHz for the USB. In either case, the exciter oscillator frequency is combined with the 7.8 MHz so that the carrier will be above the sideband when the mode switch is set to LSB and below the sideband when set to USB. This places the exciter oscillator at the same relative frequency with respect to the sideband as was the original carrier signal. If this were not done, the audio would be reversed, when, for example, a signal originated as 2.5 kHz audio would be reproduced as 500 Hz, and vice versa.

The exciter oscillator frequency is very special in several respects. For one, its amplitude must be more than five times that of the IF —possibly up to 20 times greater than the voltage from the IF stages. The exciter oscillator frequency must also be precisely 1500 Hz above the IF when receiving LSB and 1500 Hz below the IF when the mode switch is set in the USB mode. This is precisely the relationship of the original carrier to the center frequency of its sidebands. Intelligibility is greatly impaired if the exciter oscillator frequencies are as little as 10 Hz away from their desired frequencies. Clarifiers may be used to adjust the exciter oscillator frequency accurately.

The product detector combining the exciter oscillator and IF signals is the one circuit peculiar to SSB receivers. It is shown in Fig. 6-10. The frequency from the exciter oscillator and the intermediate frequency are combined in the diode bridge consisting of D1, D2, R3, and

FIG. 6-10 Product detector.

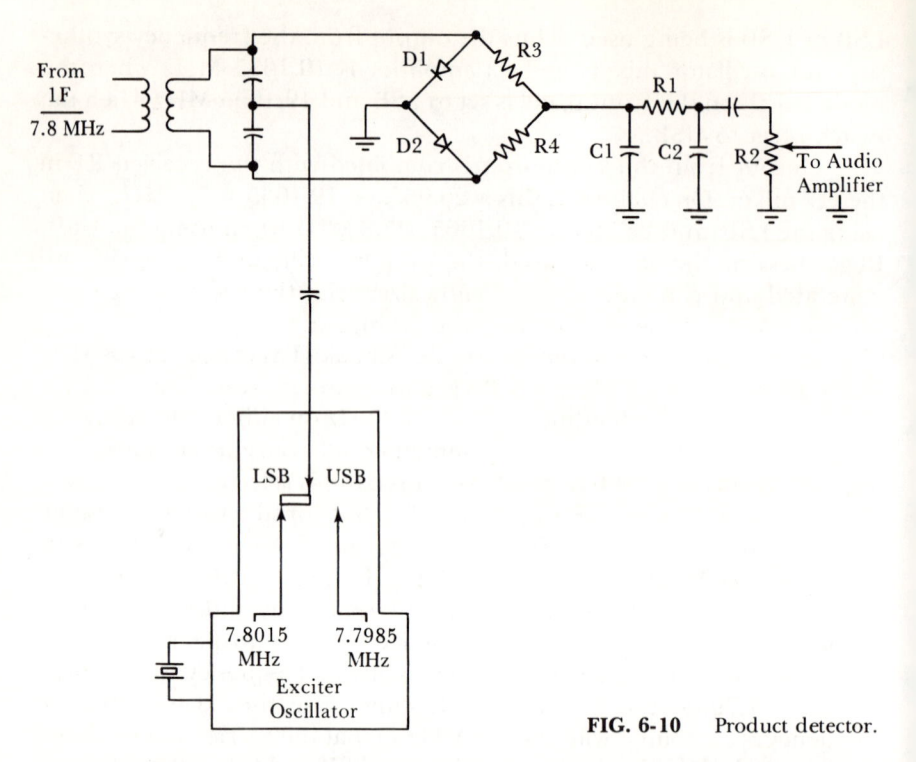

R4. The combined signal is also detected in this bridge. Combinations of audio and RF are at the output of the bridge. RF is filtered by the C1, R1, C2 circuit so that only audio remains across C2 for application to level control R2. Output from R2 is fed to the audio amplifier. There is, of course, no output from the product detector when there is no IF present.

Another method of detecting SSB signals involves combining the exciter oscillator frequency and IF into a conventional AM-type signal. It is then detected by methods described earlier in this chapter, using a single diode. Despite the greater complexity of the product detector, it is preferred to the single diode circuit because it has the distinct advantage of minimizing harmonic and intermodulation distortion components in the output.

7

Troubleshooting the Complete Transceiver

Stages in the CB radio were dissected and analyzed in the previous two chapters. However, a transceiver is composed of many stages and is an integrated whole. Unlike Humpty-Dumpty, the pieces can and must be reassembled to form a useful entity. This we shall now do.

Dual-Conversion Transceiver

The Puma 23C bearing the Pearce-Simpson label, is a transceiver composed of the stages described in detail in earlier chapters. We shall use the schematic shown in Fig. 7-1 to trace easily many of its circuits.

In the receive mode of operation, signal from the antenna jack, J1 ANT, is fed through coils in the various matching and isolating networks to RF transformer T1. After being amplified by transistor TR1, the received signal is transformer coupled to the HF mixer, TR2. One of six

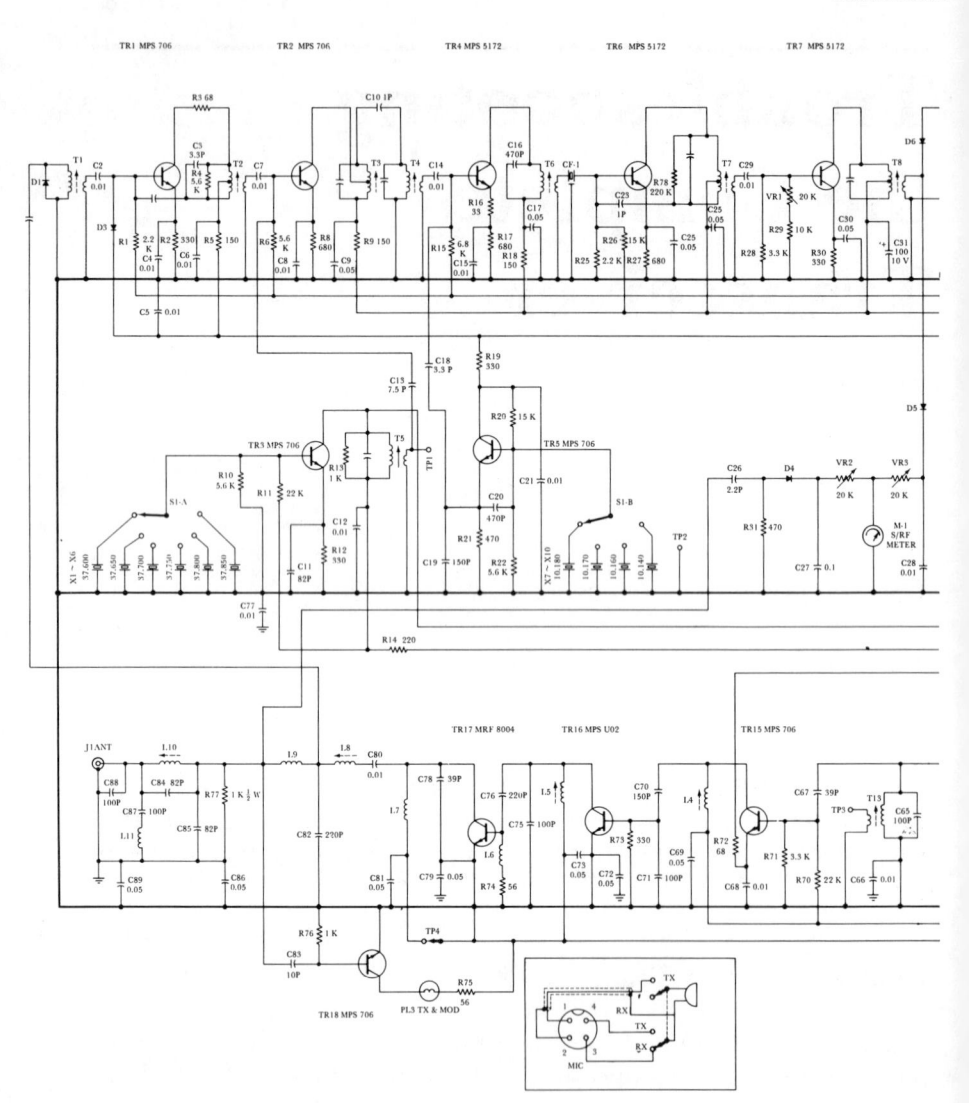

FIG. 7-1 Puma 23C transceiver. (Courtesy Pearce-Simpson, Division of Gladding Corp.)

Dual-Conversion Transceiver

DIAGRAM

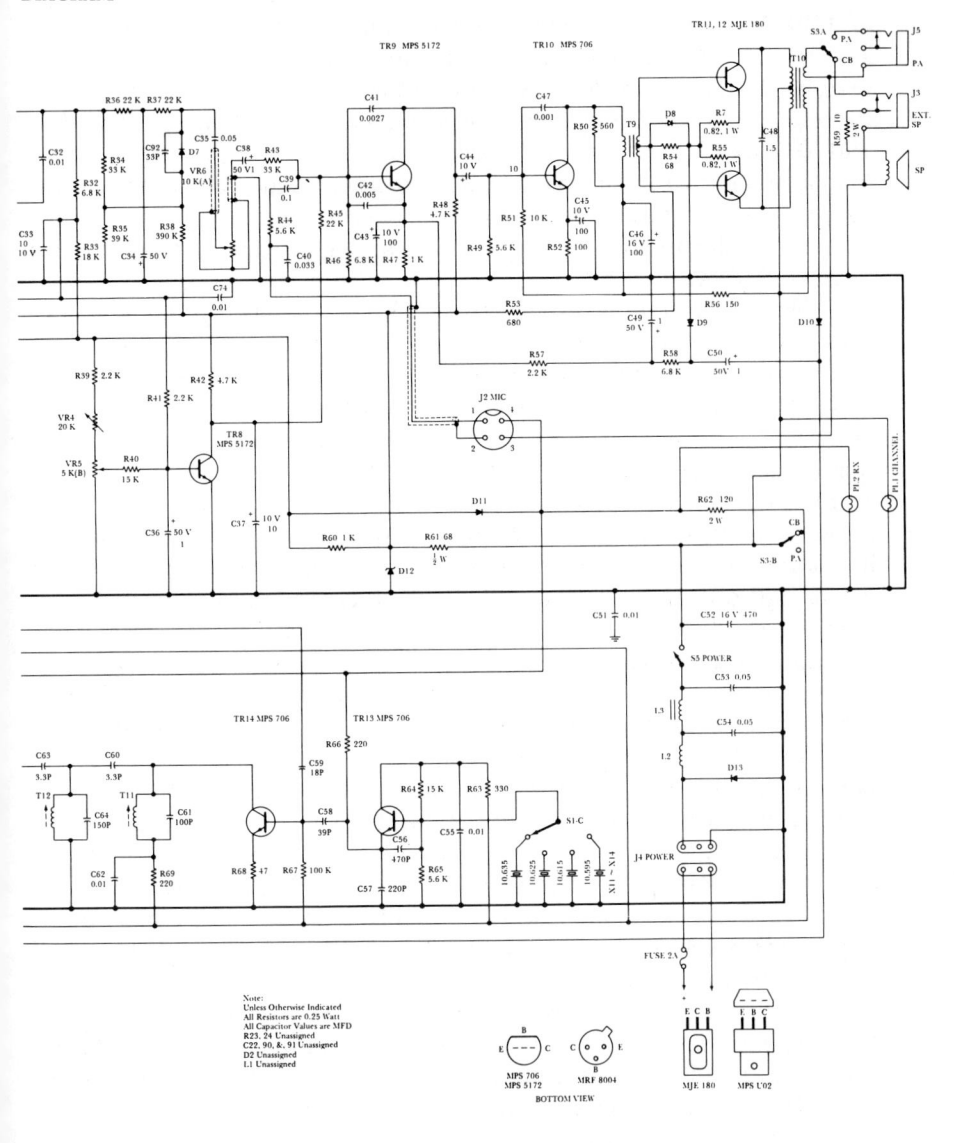

frequencies from the crystal-controlled HF local oscillator, TR3, is coupled through capacitor C13 to TR2, where it mixes with the amplified RF from the antenna. The high IF is formed here. Transformers T3 and T4, adjusted to pass only high IF signals, couple them to the LF mixer, TR4. A selection of four frequencies from TR5 is made available and fed to TR4. Here it is combined with the high IF to produce the 455-kHz low IF. The 455-kHz output from TR4 is filtered and amplified by several IF stages as well as by crystal filter CF1. Signal from the IFs is detected by D6 (noise is limited by the circuit around D7; see Fig. 6-4) and fed through audio amplifier stages TR9 through TR12 to the loudspeaker for reproduction. TR8 acts as the squelch gate transistor, using the circuit described in Chapter 6 and shown in Fig. 6-6. AGC is taken from the junction of R32 and R33 and applied to transistors TR1, TR2, and TR4.

When transmitting, signal from TR3 is combined in mixer TR14 with one of the four frequencies available from transmit oscillator TR13. The various combinations form the RF for eventual transmission. Buffer and driver stages TR15 and TR16 conduct the signal to RF power output transistor TR17. In the meantime, the audio from the microphone (pin 1 on the jack labeled J2 MIC) is fed to the base circuit of audio amplifier TR9. It is amplified by the balance of the audio circuit. The lower of the two secondary windings on the output transformer is placed in a series arrangement between the power supply and the collectors of driver and output transistors TR16 and TR17, so the RFs in these circuits are modulated by the audio. The modulated RF is fed to the antenna jack for radiation.

Now let us look at the components in the circuit used for switching the transceiver from the receive to the transmit mode of operation. In the schematic of the microphone (at the bottom of Fig. 7-1), a double-pole double-throw switch is shown. The upper pole is used for connecting the microphone cartridge to the amplifier when the switch is set to TX (transmit). The hot lead of the cartridge is connected to pin 1 of the plug labeled MIC. When the plug is in jack J2 MIC, the hot lead of the cartridge is connected to the base of TR9 through resistor R44 and capacitor C39. The second lead from the cartridge is connected to pin 2 in the microphone plug. It is connected to the circuit ground in the transceiver through its corresponding pin in jack J2 MIC.

The lower pole of the switch performs the function of connecting pin 3 on the microphone plug to pin 2 (and to circuit ground through jack J2) in the RX (receive) setting and to pin 4 in the TX position of the switch. Tracing the circuit from jack J2, the speaker is returned to ground through pin 3 only in the receive mode of operation. Completing the connection through pin 4, transmitter oscillator transistor TR13 and the ensuing mixer and buffer stages, TR14 and TR15, are all returned to ground and consequently made operative only in the transmit mode.

The basic transceiver circuit is evidently all there. So are some gadgets. For example, there is a meter M1 between oscillator transistor TR5 and squelch gate transistor TR8. It is connected to neither transistor. M1 is linked to the final IF transformer T8 of the receiver through meter rectifier D5 and sensitivity control VR3. When receiving, the pointer on M1 indicates the strength of the received signal in S units. In this transceiver, the pointer points to S9 when a signal of 100 μV is being received. Each S unit below S9 denotes that the signal level is 6 dB less, or one half the signal at the next higher marking. Thus, if the meter pointer indicates S8, the input signal is 50 μV, or 25 μV if the pointer is at S7, and so on. Similarly, when transmitting, the meter is connected to the antenna circuit through rectifier D4 and adjustment control VR2 to indicate output power. Shunting capacitors C27 and C28 filter the RF and IF, respectively, so that these frequencies are not applied to the meter coil.

Bulbs PL1 through PL3 are also useful, functional additions to this CB radio. Whereas PL2 lights to indicate that the transceiver is on, PL1 lights only when the push-to-talk switch in the microphone is in the transmit mode. The circuit involving PL3 was discussed in Chapter 5. Light from this bulb fluctuates in brilliance as the level of the audio modulating signal varies.

Each gadget is useful. Other gadgets and circuits used to accomplish the functions noted here have been described, or at least mentioned, in previous chapters. Once you are familiar with the intricacies of the circuit details summarized here, it is a trivial task to analyze more complex transceivers using other circuits and features. Even SSB transceivers should not faze anyone once the peculiarities described in Chapters 2, 5, and 6 are recognized.

Practical Troubleshooting Procedures

Before we begin to describe methods that should be used when troubleshooting the Puma 23, it should be pointed out that the discussion centers around this particular transceiver for several basic reasons. As a starter, the diagram is well drawn and can be readily used in our discussion. The circuit itself is straightforward and thus lends itself to ease of analysis. It should not be inferred that this transceiver has any tendency to be troublesome. It compares favorably in performance and durability with other units on the market.

When servicing a transceiver in the service shop, first determine the problem. Check if the transceiver is operating within reason in the receive mode. Connect a dummy load to the antenna input. Turn the volume control up to its maximum clockwise setting. Set the squelch control to the end where it has no effect in limiting the noise. Drape the output leads from an RF signal generator (with the frequency monitored by a digital counter) around the load. Adjust the generator so that it supplies a modulated 26.965 MHz, and set the transceiver channel selector switch to channel 1. If all is well, the modulating frequency is heard through the loudspeaker. Repeat this test for all other channels using the frequencies designated in Table 1-1. Note if the output on any channel or channels is weak or nonexistent. If you cannot get reception on all channels, check individual sections of the receiver more carefully. Should there be no output on only one or on a few of the channels, the problem is either with the selector switch or with one of the crystals in the synthesizer.

You can usually determine which crystal is bad from the numbers of the channels not operating properly. For example, if the 37.6-MHz crystal is defective, outputs on channels 1 through 4 are impaired; if the 10.18-MHz crystal is defective, outputs are not proper on channels 1, 5, 9, 13, 17, and 21. You can use Fig. 5-2 to determine which crystals are associated with the various channels in 23-channel single-conversion transceivers. Use Table 7-1 for ascertaining the crystals applied when set to the different channels in the dual-conversion Puma 23C.

Should signal fail to pass through the entire circuit, conventional troubleshooting methods are usually sufficient for locating not only the defective stage, but also the inoperative component in that stage. A good procedure always starts with voltage checks. You must first ascertain if the supply voltage is reaching the audio output stages. Then measure voltages at the various transistor elements; start at the push–pull output devices and work your way back to the RF amplifier. If the information is available, compare the measured voltages with standard values supplied by the manufacturer. If they differ by much at any one stage, it is an immediate indication of where to check further for the problem component. Should you be working without voltage information from the manufacturer, you can usually judge by yourself what the voltages should be. You know the voltage at the supply. If this voltage is applied through a transformer, the voltage at the collector is very close to that at the supply. Just what the relative voltages should be at the other terminals of the transistors and diodes can be deduced from the discussion in Chapter 4.

Resistor measurements should be performed on the various components in the circuit. Check for open coils in the loudspeaker as well as

Practical Troubleshooting Procedures 157

Table 7-1   Receiver Crystals (MHz) Used When Set to Different Channels

Channel	High-Frequency Crystal	Low-Frequency Crystal	Channel	High-Frequency Crystal	Low-Frequency Crystal
1	37.6	10.18	13	37.75	10.18
2	37.6	10.17	14	37.75	10.17
3	37.6	10.16	15	37.75	10.16
4	37.6	10.14	16	37.75	10.14
5	37.65	10.18	17	37.8	10.18
6	37.65	10.17	18	37.8	10.17
7	37.65	10.16	19	37.8	10.16
8	37.65	10.14	20	37.8	10.14
9	37.7	10.18	21	37.85	10.18
10	37.7	10.17	22	37.85	10.17
11	37.7	10.16	23	37.85	10.14
12	37.7	10.14			

in the audio, IF, and RF transformers. Next reassure yourself that the various capacitors do not present low resistances or shorts to the circuit. Once these tests have been completed, signal-tracing methods should be employed.

After having assured yourself that the oscillator circuits around TR3 and TR5 are operating at the required frequencies on all channels, proceed to trace the signal from the RF input to the audio output, using a signal tracer or oscilloscope described in Chapter 4. Check for the presence of signal as well as for the amplitudes of the signal at the various stages. Once you have located a stage that will not pass signal, you must resort to ohmmeter and voltmeter measuring procedures to ascertain which component in the circuit is the culprit. The details supplied in Chapters 5 and 6 should be helpful in this task.

Signal-tracing procedures as well as voltage-measuring troubleshooting methods are not infallible. Should you require further analysis, signal-injection techniques may be tried. Feed audio signal at about 1 kHz into one audio stage at a time, starting at the audio output and working your way back to the detector. If signal passes properly through these stages, then inject modulated IF and RF signals at the proper frequencies into earlier stages. Once a stage that fails to pass the signal is located, measure resistances and voltages carefully around that stage. You should be able to locate the defective component even if you must check each one individually in the affected circuit.

Don't forget the antenna coil preceding the first RF stage. The defect may be right in the antenna jack or components in series with or across the antenna proper.

There are a number of items that you should suspect when different symptoms are evident.

Defects in the Receiver Section

Hum Through the Loudspeaker. Obviously, hum cannot originate at a dc voltage source. Only base stations powered by alternating current are primary objects of this defect. Hum runs rampant through the circuit if residual ripple has been allowed to pass through the power supply filter circuit. It will be entirely intolerable if it is applied from the voltage source to the base circuits of the various stages. Because hum frequencies are within the audible band at 60 and 120 Hz, it is obvious that ripple wreaks havoc if it sneaks into audio amplifier circuits. Ripple is likewise undesirable in RF and IF amplifier voltage supplies, for signals here can be modulated by hum frequencies, which will appear at the audio output after being detected.

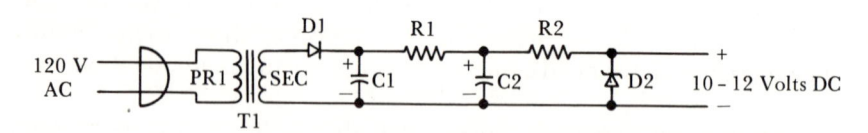

FIG. 7-2 Half wave power supply circuits.

A basic half-wave power supply circuit is shown in Fig. 7-2. The turns ratio of transformer T1 is designed to reduce the 120-V ac at its primary to a lower voltage at the secondary. The magnitude of the secondary voltage is not the same for all transceivers. It also depends upon the actual voltage at the primary of T1 at a particular instant. Voltage at the secondary of the transformer is rectified by diode D1. Positive 60-Hz pulses are at the cathode of the diode. Filtered by the C1–R1–C2 network to reduce ripple, a reasonably smooth direct current is applied through R2 to zener diode D2. Voltage is held constant across the diode at its breakdown voltage, usually somewhere between 10 and 12 V. The diode serves two functions. One is to set and maintain a constant supply voltage for the various oscillator, mixer, and amplifier circuits. Zener diode D2 is an important factor in holding the supply voltage constant despite line and load variations. The second function of D2 is to add to the effectiveness of the filter circuit in reducing ripple by acting in conjunction with resistor R2.

Hum is usually due to a defective capacitor in the filter circuit. It

Practical Troubleshooting Procedures 159

can best be checked by jumping a good capacitor across the suspected capacitor in the circuit and noting if ripple is reduced when the substituted part is used. If C1 is defective, connecting another capacitor of like value across it raises the voltage at the junction of D1 and R1.

A leaky rectifier D1 can likewise cause excess ripple. It should be checked by noting its forward and reverse resistance on an ohmmeter, as discussed earlier. Should this rectifier require replacement, use a component that has voltage breakdown and current ratings equal to or better than those of the device specified by the manufacturer of the equipment. Should this information not be available, a 200-V device rated to pass a constant 3 A without heat sinking should be used. If there is a choice, select the rectifier with the highest forward surge current rating and with a specified breakdown voltage greater than three times the voltage that is supposed to be across C1.

Hum, high voltage, and poor regulation can be due to a defective zener diode D2. Use the ohmmeter method or, preferably, a transistor tester to check this device. If it is defective, be sure the breakdown voltage and reverse current (or power) ratings of the replacement are as required.

Somewhat more effective power supply circuits are shown in Fig. 7-3. Full-wave rectification is achieved in both arrangements. In Fig.

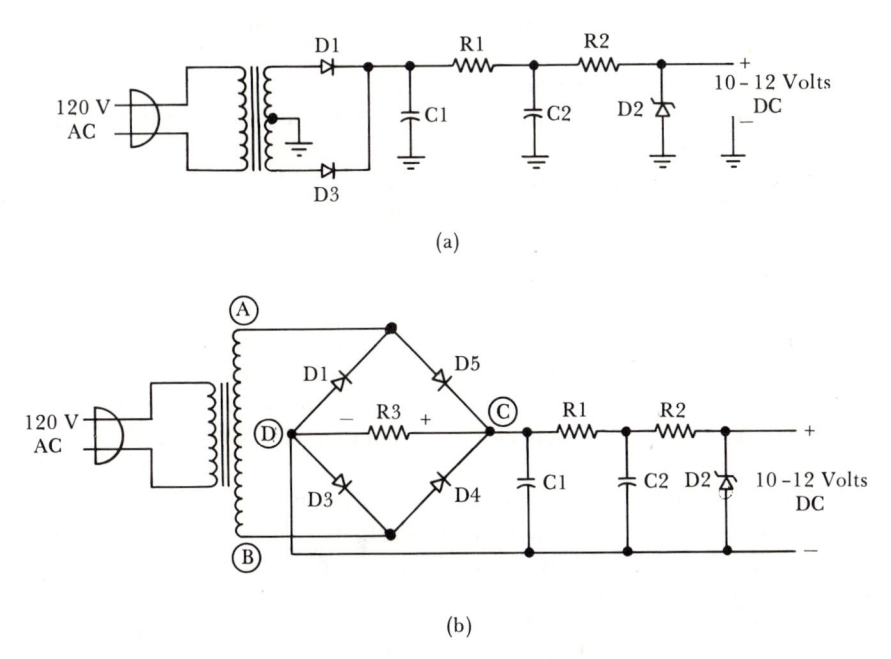

(a)

(b)

FIG. 7-3 Full wave power supply circuits.

7-3a, D1 conducts during one-half of the cycle; D3 conducts during the alternate half. Positive voltage is at the junction of both diodes with C1. Because there is conduction during both halves of the cycle, the fundamental ripple frequency at the output of this circuit is 120 Hz. This frequency is easier to filter than the 60-Hz ripple from the half-wave rectifier circuit. Even if the components in the filter circuit are the same as those used in the circuit in Fig. 7-2, ripple here is considerably lower.

The bridge circuit in Fig. 7-3b also supplies dc voltage with a fundamental 120-Hz ripple. During the portion of the cycle when terminal A of the transformer is positive with respect to B, there is a complete circuit through D5, R3, and D3. During the alternate half of the cycle, conduction is through D4, R3, and D1. Terminal C of R3 is positive with respect to D during both halves of the cycle. Voltage from R3 is applied in the proper polarity to the filter and regulator circuits.

Defects that can occur in these circuits are not unlike those described for the half-wave rectifier. About the only difference is that more rectifiers must be tested here than in the previous type of circuit, if there should be operative failure.

Feedback regulator circuits can be used to stabilize the output voltage. Better regulation and lower hum levels are achieved when using this type of circuit than when using only a zener diode as the regulating device. A basic regulator circuit is shown in Fig. 7-4. Most transceivers use more complex feedback regulators, but the circuit in each case is basically as described below.

Filtered dc voltage across C2 is fed to the collector of transistor Q1. When it conducts, dc voltage is also at its emitter. Q1 conducts only if current through R6 is allowed to pass into its base–emitter circuit to turn it on. Should Q2 conduct too heavily, all current from R6 flows through its collector, and no current remains for the base of Q1. Q1 is consequently cut off. In the meantime, zener diode D2 is biased into the breakdown region by voltage applied to it through R7 and R8.

With formalities out of the way, let us assume that some current from R6 flows into both the collector of Q2 and the base of Q1. There is voltage at the output of Q1 and across resistors R4 and R5. The base of Q2 is at the junction of R4 and R5. Base–emitter current flows into Q2 owing to the difference between the voltage at the R4–R5 junction and the voltage at the cathode of D2. The base current, multiplied by the beta of Q2, is the current through the collector of that transistor. Current is supplied to the collector through R6. There is also sufficient current left in R6 to flow through the base–emitter junction of Q1, turning it on. The base–emitter current in Q1 should be of such value that a regulated voltage of 10 or 12 V is at its emitter, which is in turn divided by resistors R4 and R5 and applied to the base of Q2. Here the circle is repeated through the entire feedback loop. Once the values of

Practical Troubleshooting Procedures 161

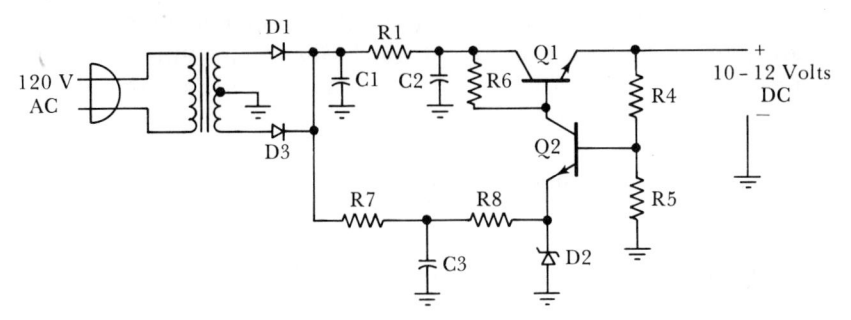

FIG. 7-4 Feedback regulator.

R4 and R5 are chosen, voltage at the emitter of Q1 is maintained constant. The ratio of R4 to R5 is important in setting the output voltage.

Voltage across zener diode D2 is constant at its breakdown potential. If Q2 is a silicon transistor, there is a fixed 0.7 V from its base to its emitter. As D2 and the base–emitter junction of Q2 are in a series circuit across R5, voltage across this resistor is held fixed at a value equal to the sum of the voltages across the two junctions. A change in the relative sizes of resistors R4 and R5 changes the base current through Q2 and consequently affects the voltage at the output of the regulating circuit.

You will be unable to set the output voltage if either transistor is defective. Should the voltage change slightly, it can be reset by readjusting the value of either resistor R4, resistor R5, or both. If changing these components does not affect the output voltage, check all semiconductors. First note if the voltage across D2 is as specified. Then check Q1 for a collector–emitter short. Finally, Q2 may have been destroyed and should be checked in circuit using a transistor tester.

As before, hum is a function of the condition of components in the filter circuit, especially capacitors C1 and C2. Voltage across D2 should be relatively free of ripple. This is assured by the presence of the network involving R7, R8, and C3.

Up to now we have associated hum only with the power supply. We have even diverted somewhat from the discussion in order to describe ac supplies used in base stations. Hum can also be due to other factors. If a coupling capacitor in the audio circuit used to connect the collector of one transistor to the base of the next one turns leaky, it may turn into an undesirable path for power supply ripple. An open level control can similarly be the cause of hum problems. Mobile and base stations are both susceptible to hum from this defect, for hum can be induced into the circuit from ac fields generated by nearly equipment or power lines.

Intermittent Reception. This condition may be generated by a defective capacitor in the filter circuit of the power supply. Here connections from the capacitor leads to the plates inside the component may be poor. Actually, any intermittent components (including relays and crystals) or poorly soldered connections can produce an erratic output condition. A defect of this type can best be located by jiggling the wires and components in the circuit. As far as the crystal is concerned, the substitution method is the most desirable procedure to ascertain if the component is really bad.

Should operation be erratic when switching from channel to channel or when jiggling the shaft on the switch, the switch is probably the defective component. First check all connections to the switch. If leads and components are soldered solidly to the lugs, spray the switch contacts with a tuner cleaning fluid. If nothing helps, the switch must, of course, be replaced.

No Output. Before tackling this type of defect, double-check the performance of the transceiver to be certain that this condition does exist. Connect the antenna to its jack. Turn the level control fully clockwise and the squelch control to the setting where it will not limit the output signal. Be sure the PA–CB switch is in the CB position. Now if there is no audio at the output, you can be pretty sure that the symptom is real.

Defects of this type are usually easier to handle than intermittents. Use the voltage-measuring, signal-tracing, and signal-injection procedures described above. Should there be no supply voltage, no bulbs will light. Check the fuse. Bulbs may not light and the fuse can blow if connections are reversed from the CB radio to the dc power supply. Diode D13 in Fig. 7-1 protects the transceiver from a mishap should connections to the supply be inadvertently reversed, but the fuse will blow.

Should the defect be in the audio circuit, first check the relay, speaker, and output transformer. Do not forget that no audio output may be due to a defect in the noise limiter and squelch circuits.

Defective coupling capacitors can prevent signal from passing through IF and RF circuits as well as through the audio amplifiers. But in the case of IFs and RFs, open coils and transformers are frequently the cause of no output. Furthermore, in HF circuits, open capacitors are more likely to provide a symptom of weak signal at the output rather than no signal at all.

Weak Signal. There is hardly a need to mention here that a defective antenna or relay contact will radically affect the sensitivity as well as other characteristics of the receive section in the CB radio. Poor sensitivity can usually be cured by carefully realigning all RF and IF stages. A somewhat inactive oscillator crystal or an open bypass capacitor can also

Practical Troubleshooting Procedures 163

produce this symptom. In Fig. 7-1, check C4, C6, C8, C9, and the other capacitors in all similar circuit locations.

Noisy Background and Weak or Entirely Nonexistent Signal. Once again the antenna and relay contacts are the biggest offenders. Also check the high-frequency and receiver crystals as well as the channel selector switch. Proper alignment of the entire receiver, especially the RF stage, is essential if this condition is to be relieved.

Distorted Audio. The factor producing this symptom can best be located by using signal-tracing procedures. A defective coupling or bypass capacitor is the most likely cause. A crystal not on frequency or a poorly aligned IF stage may also be a factor leading to distorted audio at the output. Transistors and diodes in any of the circuits may go bad, but do not forget the resistors, which can cause significant disorders if they should change in value. Before replacing a semiconductor device, check all resistors in the circuit with an ohmmeter.

Poor Selectivity. Channels may overlap in one or more settings of the channel-selector switch. If the design is proper, this symptom is usually due to a misaligned IF transformer.

Inoperative Squelch Control. If this defect should develop, check the circuitry around the squelch gating transistor TR8, as well as the device itself. Variable controls are subject to failure, so be sure to check if VR4 and VR5 are in proper shape.

Defects in the Transmitter Section

Just as there are signs to help pinpoint defects in the receiver section of the transceiver, symptoms can also be used to indicate the inoperative or poor components in the transmitter section. An RF wattmeter, dummy load, digital counter, and modulation meter are the instruments required in this instance. Before proceeding with detailed analysis, use a digital counter to check if all channels are being transmitted on their assigned carrier frequencies. You can determine which crystals are defective from the numbers of the channels that are not on frequency. Channels affected by signals from specific HF crystals are shown in Table 7-1. A listing of transmitter crystals used for the different channels in the Puma 23C is in Table 7-2. Remember that these frequencies apply to the transceiver in Fig. 7-1 and to many other CB radios, but the schedule is by no means universal.

Now to troubleshooting by symptoms; we shall assume that the oscillators are in working order.

Table 7-2   **Transmit Crystals Used When Set to the Different Channels**

Channels	Crystal Frequency (MHz)
1, 5, 9, 13, 17, 21	10.635
2, 6, 10, 14, 18, 22	10.625
3, 7, 11, 15, 19	10.615
4, 8, 12, 16, 20, 23	10.595

Carrier Modulated by Hum. See the discussion of power supply hum in the section on receiver troubleshooting by symptoms.

No Transmission. Push-to-talk switches in microphones are good for a predetermined number of operations. Moisture may attack the stators and wipers in the switch, wreaking havoc on contacts between them. So if there is no RF output noted on the power meter (connected between a dummy load and the antenna jack on the transceiver), your prime suspect should be the push-to-talk switch in the microphone. While checking the microphone, also inspect the cable and the connector at the end of that cable. A broken lead here can have the same effect as an open switch.

Should your CB radio use a relay in the switching circuit, corroded or pitted contacts here will not allow the transceiver to go into the transmit mode.

Of course, faults may have developed in the transmitter circuit. Assuming that oscillators are operating properly and are on frequency, the defect may lie anywhere between TR14 and TR17 in the unit shown in Fig. 7-1. Although most failures are actually in the output transistor (TR17 in this circuit), a dip meter should be used to check if proper RF signal is at each of the inductors in the circuit. A lack of signal at any inductor is a pretty good indication that the transistor preceding that component is defective.

For modulation purposes, dc voltage is applied to the collectors of transistors TR16 and TR17 through one winding of output transformer T10. If there is no voltage at both of these collectors, check the applicable winding of the transformer and diode D10 for continuity.

Low Output. If the transmit and modulation indicator bulb PL3 does not light, you immediately have a good indication that the output from the transmitter section is low. TR17 has probably been damaged by a short in the antenna cable or connector. Because the RF power output transistor is expensive, check it in a transistor tester before deciding that the device must be replaced. If it should prove to be in good operating

Practical Troubleshooting Procedures 165

condition, test coupling capacitors C67, C70, C76, and C80, by jumping each one with a known good one. If the proper power output level is restored by jumping any one of them, you have located the defective component.

Low or No Modulation. Assuming that the receiver works properly, check the applicable secondary winding of the output transformer T10 as well as the associated diode. Tests can be performed with an ohmmeter, or use an oscilloscope, to note if there is any audio across this winding when the transceiver is in the receive mode of operation.

The fault may be with the microphone, for a short may have developed in the cable between the "hot" audio lead and its shield in the connector from the microphone to the transceiver, or at the cartridge. Cartridges are prone to failure. They should be checked on an oscilloscope by whistling into the microphone and noting the output level on the screen.

In the circuit in Fig. 7-1, TR9 serves as an amplifier for the received signal as well as for the output from the microphone. In transceivers using this arrangement, good reception is an indication that the transistor is operating properly. Many transceivers on the market contain a separate microphone preamplifier stage. Should one of these amplifier stages be defective, modulation will be adversely affected.

Distorted Modulation. This is usually due to a defect in the microphone or audio amplifier circuit. A leaky capacitor or partially shorted modulation transformer are prime suspects. Should the audio circuit be shipshape, then you must test the RF section of the transmitter section very carefully. Primarily, be sure that all circuits here are tuned properly.

You should check the modulated signal for distortion on a wideband oscilloscope. If you do not have an oscilloscope that can display signals in the 27-MHz range, check distortion by transmitting, using the transceiver under test, and observing the quality of the received signal on a second CB radio. Should you use the two-transceiver method, separate the transmitter and receiver by at least 1 mile. If they are in close proximity, the received signal can be sufficient to overload the input stage of the receiver. Consequently, the reproduced signal will be distorted even though the transmitted modulated RF is clean.

Inoperative Automatic Modulation Limiting. If you can modulate the RF more than 100 percent, the limiting circuit is not operating properly. In Fig. 7-1, this can be due to components at the modulator winding of audio transformer T10, such as C50, D9, or R58, as well as to a short in the emitter circuit of TR9.

Public-Address Operation

The inclusion of a public-address-system (PA) facility in just about every CB radio seems unjustified, as no one seems to use it. But it is an advertising feature and costs almost nothing to build into the transceiver.

By setting the switch on the front panel to PA and the push-to-talk switch on the microphone to the transmit position, the transmitter circuit is disabled while signal from the microphone is magnified by the audio amplifier. After amplification, the audio is applied to a loudspeaker connected to a special PA speaker jack on the rear apron of the transceiver. In Fig. 7-1, the two poles on switch S3 are all that are required to convert the transceiver into a PA amplifier.

First note switch section S3B. When set to the CB position, voltage is applied through this pole to all oscillators, as well as to the mixer and predriver stage of the transmitter section. Once switched to PA, voltage is removed from all these circuits. It is applied through other paths to the audio amplifier consisting of TR9 through TR12.

After the signal from the microphone has passed through these stages, it appears at the secondaries of audio output transformer T10. The applicable speaker output jack is selected through switch section S3A. In the PA setting, jack J5 is selected. A loudspeaker at this jack is used to reproduce audio originating at the microphone.

Troubleshooting procedures do not differ radically from those described above. Because of the overabundance of leads from the microphone case to the CB radio proper, much consideration must be given to them when tracing for any defects.

An Innovation

While driving, it is inconvenient to adjust the controls on the panel of the transceiver, unless it is close at hand. Loudspeakers in transceivers mounted in automobiles are oriented to face the floor of the car. Consequently, audio is somewhat restricted. The transceiver in Fig. 7-5 is designed to overcome many of these drawbacks.

The mobile transceiver can be mounted at any out-of-the-way location in the car or truck. Only access to the microphone is required by the operator. All important controls are built into the microphone case, including the channel selector switch, volume and squelch potentiometers, and the push-to-talk button. In addition, one transducer is used both as the microphone cartridge and loudspeaker, so you can listen to conversations and talk into the same small hand-held package.

Troubleshooting procedures do not differ radically from those

Practical Troubleshooting Procedures 167

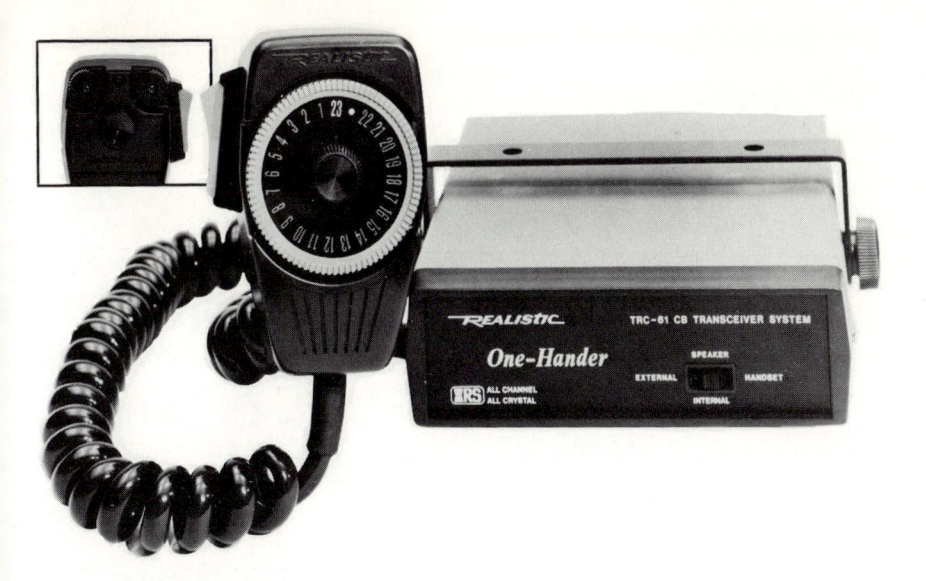

FIG. 7-5 An innovative design in transceivers. (Courtesy Radio Shack.)

described above. Because of the overabundance of leads from the microphone case to the CB radio proper, much consideration must be given to them when tracing for any defects.

8

Choosing a Microphone

A microphone is supplied with just about every transceiver sold today. No freedom of choice is extended to the customer as to which microphone he acquires with his purchase. Even when it must be replaced, the type (ceramic or dynamic) of microphone he can get is limited by the transceiver circuitry. This has been discussed in detail in Chapter 2.

Ceramic and dynamic microphones seem to be among the simplest parts of the CB rig. Actually, the physics of the transducers or cartridges and the acoustic characteristics of the housing are quite complex. Transducers are subject to failure. Other parts in the microphone are quite vulnerable and can cause problems or break down completely. High resistance may develop in the mechanical switch through use or through exposure to weather elements. A wire may break in the constantly expanding and contracting coil cord. Connections to the plug at the end of the coil cord may be severed owing to repeated tension. Any one of these breaks may impair or completely arrest the operation of the CB rig in the receive or transmit mode, or both.

Troubleshooting Microphones

Several steps must be pursued when troubleshooting any microphone, regardless of whether it is part of the original equipment or a replacement bought separately.

First, determine the switching that must be accomplished in the microphone by noting its circuit in the schematic of the CB radio involved. Then check the microphone to be certain that the switch is operating properly. Connect an ohmmeter to the appropriate pins on the plug at the end of the microphone cable. Make the tests to ascertain that the switch in the microphone makes and breaks the circuit to the various pins, as required, in both the receiver and transmit switch positions.

As an example, see the schematic of the Mura microphone in Fig. 8-1. The pins of the plug connected to the red and green leads in the cable have zero resistance between them when the switch is in the receive position and infinite resistance when the switch is set for transmitting. The reverse is true of the resistance between the red and blue leads. In both settings of the switch, resistance between pins connected to the green and blue leads must be infinite. The resistance between the yellow lead and shield on this model is infinite in the receive position and several hundred ohms (the dc resistance of the dynamic cartridge) when the switch lever is pressed for transmitting. If a ceramic cartridge is in the microphone, resistance between the pins connected to the yellow lead and the shield is infinite in both settings of the switch.

Improper ohmmeter readings can be due to a defective switch or, more likely, to a broken lead to the connector at the end of the cable. Remove the outside case from the connector and visually check the

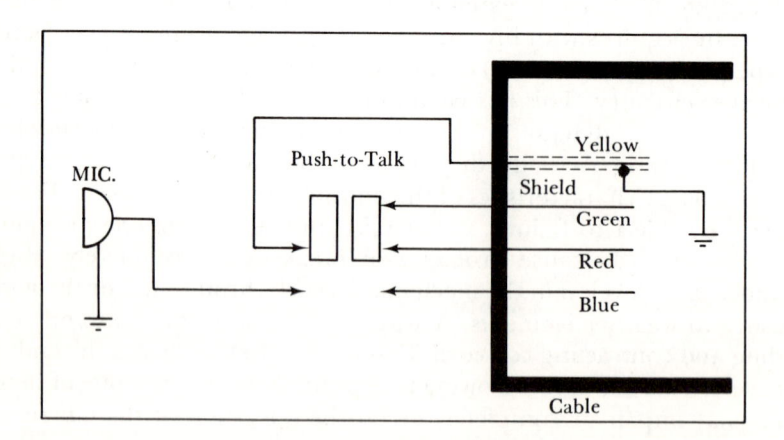

FIG. 8-1 Schematic of replacement microphone. (Courtesy Mura Corp.)

Troubleshooting Microphones 171

connections from the leads to the pins. If everything seems to be intact, carefully open the case of the microphone. Note how it was assembled, as reassembly may be tricky. Check the continuity of each lead in the cable from inside the microphone to the plug with an ohmmeter. If there is no continuity in at least one lead, look more carefully for a break. If you cannot find the break, note the colors of the leads in the cable as they are connected to specific pins in the plug. Unsolder the leads from the plug. Cut off a short length of cable at this end and resolder the leads from the cable to the plug, observing the color code. If there is still no continuity in one or more of the leads, repeat this procedure at the microphone end of the cable. Should both operations fail in restoring continuity to all leads, the cable must be replaced.

Assume that you found the cable to be in good condition, but the switch in the microphone is defective. Try cleaning it with tuner contact cleaner. If this does not help, replace it with a switch that is identical to the original. A replacement should be available from the supplier of the transceiver or microphone.

Should any part need replacement, do not remove it before noting the circuitry in the microphone and the color code of the leads from the cable as they are connected to the switch, cartridge, and pins of the plug. Mix-ups can be quite disconcerting and a big time waster.

If repairing the microphone becomes too much of a task, it should obviously be replaced with a similar unit. Unless you use some type of amplified microphone, be sure the replacement has a transducer similar to the one used in the original. Never replace an original equipment microphone that has a ceramic cartridge with a new one incorporating a dynamic transducer, and vice versa.

Connecting the leads from the cable on the replacement microphone to the proper pins on the connector may become tricky. It is a simple operation if a complete wiring guide is supplied by the manufacturer of the replacement microphone. Companies like Astatic, Mura, Telex, and Turner supply fairly complete guides. Others may do this, too, but I have never seen any. If a guide is not available, you can usually get the information by writing to the manufacturer of the microphone or by figuring it out for yourself from the schematic.

Suppose, for example, that you bought the Mura DX-115D microphone as a replacement for the one supplied with the Pearce-Simpson Puma 23C transceiver. A drawing of the microphone circuit along with plug wiring data for the Puma 23C is supplied with the microphone. Assume for this discussion that these data are not available to you and you have only the microphone schematic as in Fig. 8-1. A schematic of the Puma 23C is shown in Fig. 7-1.

A diagram of how the original equipment microphone is connected to the pins on the plug is at the bottom of the Puma 23C schematic. In the upper section (pole) of the switch in the microphone drawing, a lead from the cartridge is connected to pin 1 on the connector labeled MIC when the switch is set to the TX (transmit) position. The lead is shown inside a shield. In the Mura microphone, the cartridge is similarly connected to the shielded yellow lead through the switch. Thus the yellow audio lead from the DX-115D must be connected to pin 1 on the connector.

The cartridge of the Mura microphone is connected into the circuit (to the yellow lead and pin 1 in this transceiver) *only* when the switch is keyed into the transmit position. Different arrangements are possible. In some microphones, the cartridge is *always* connected to the audio transistor in the CB radio. It is not switched in and out of the circuit, but stays in the circuit regardless of whether the switch is in the receive or transmit mode. In the Puma 23C, the audio output from the receiver section and the cartridge are connected to the same audio stage, TR9. In the receive mode, the cartridge acts as a shunt diverting a good portion of all received audio from the input of transistor TR9. In this case, a small 10,000-Ω resistor, usually 0.25 W, is placed in series with the yellow lead. This resistor reduces the shunting effect of the cartridge. This same solution can be used if the receiver has a tendency to oscillate in the receive mode owing to shunting by the cartridge in the replacement microphone. However, now the resistor may have to be increased to some value between 12,000 and 47,000 Ω.

The shield, the second lead from the cartridge, and the wiper on the lower pole of the switch are all shown connected to pin 2 on the MIC plug of the original microphone. Following this, connect the shield from the Mura microphone to pin 2 in the plug. Also connect the red lead (connected to the wiper on the second pole of the switch in the replacement microphone) to this same pin 2 on the plug.

In the RX (receive) setting of the switch, the wiper on the bottom pole is connected through the switch to pin 3 in the plug. The wiper is connected to pin 4 of the plug in the TX setting of the switch. Corresponding leads in the replacement microphone are green and blue. Thus the green lead should be connected to pin 3 in the plug and the blue lead to pin 4. If connections are reversed to these two pins, the transceiver is keyed on when the switch on the microphone is in the receive position and you can receive when it is set to transmit. Should this occur, RF will be generated for transmission in the receive setting of the switch, but modulation will not be possible. To correct this situation, you need only reverse the green and blue leads.

All leads from the replacement microphone are usually used when it is connected to a transceiver designed for electronic switching (see Chapter 2). Less wiring is usually required from the microphone when it is used with CB radios featuring relay switching. In this latter type of rig, the green lead of the microphone in Fig. 8-1 is frequently not used and can be cut off at the connector end of the cable.

Replacement Microphones

Circuits in both dynamic and ceramic replacement microphones can be as shown in Fig. 8-1. The Mura DX-115 in Fig. 8-2 can be purchased with either type of cartridge. A 500-Ω dynamic cartridge is used as the transducer in the DX-115D; the DX-115C features a ceramic transducer. Both microphones can be used with transceivers that employ either relay or electronic switching.

FIG. 8-2 Replacement microphone using dynamic cartridge. Model DX-115-D. (Courtesy Mura Corp.)

Turner's model 450C in Fig. 8-3 has a ceramic cartridge. It is wired for use with CB radios featuring relay switching. A companion version, the J450C, must be used on transceivers with electronic switching circuits.

All the microphones described thus far can be held conveniently in the hand when transmitting. Hence they can be used with both mobile and base stations. Microphones designed specially for use with base stations can be placed on a table. They do not have to be held. Only a bar must be pressed when transmitting. One microphone of this type, the Turner 254HC shown in Fig. 8-4, features a ceramic cartridge and is usable with transceivers featuring either relay or electronic switching.

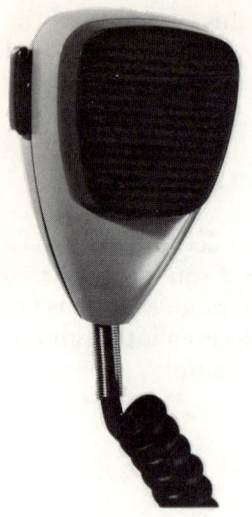

FIG. 8-3 Replacement microphone using ceramic cartridge. Model 450C. (Courtesy Turner Division, Conrac Corp.)

FIG. 8-4 Base station microphone using ceramic cartridge. Model 254 HC. (Courtesy Turner Division, Conrac Corp.)

FIG. 8-5 Noise cancelling microphone using two dynamic cartridges. Model DX-119. (Courtesy Mura Corp.)

The next step up the replacement ladder is the noise canceling microphone. It is especially useful where background or ambient noise is high and transmission is marred or entirely impossible.

Replacement Microphones

Noise cancellation can be accomplished in different ways. The Mura DX-119 in Fig. 8-5 uses two dynamic cartridges. They are separated by several inches in the microphone housing, but face in the same direction. Both cartridges receive the same ambient noise signal. The cartridges are connected in parallel. They are connected in such a manner that the noise at the output of one cartridge is out of phase with the noise at the output of the second cartridge. Noise is canceled at the output of this combination. But speech is impressed on only one of the cartridges when the microphone is held near the mouth. There is no discernable cancellation of this audio as no (or very little) speech reaches the second cartridge. Desirable audio intelligence passes through this microphone as easily as it passes through the more conventional type of replacement microphone.

A single dynamic cartridge is used to accomplish this goal in the Turner NC350DM in Fig. 8-6. Here noise signals are directed to the front and rear of the cartridge element, so that the total effect of ambient noise on the diaphragm is nil. You then speak close to the front of the cartridge so that no or very little speech will stray to the rear. Speech will not be canceled, but the noise will.

It may be difficult to hear received signals in high noise areas. A headphone can help in this situation. It can be connected to the transceiver. Jack J3 in Fig. 7-1 is supplied for this purpose. When the plug from a headphone is inserted into the jack, the circuit to the loudspeaker in the CB radio is opened. Only the headphone is connected to the output of the audio circuit to reproduce the received signal.

Convenience can be increased by mounting a microphone onto the headset. With this combination, your hands are relatively free when transmitting and receiving. The Mura CBX-111 shown in Fig. 8-7 is one such arrangement using a dynamic cartridge. As an added bonus, the

FIG. 8-6 Noise cancelling microphone using a single dynamic cartridge. Model NC 350 DM. (Courtesy Turner Division, Conrac Corp.)

Choosing a Microphone

FIG. 8-7 Boom microphone headset using noise cancelling dynamic cartridge. Phone level is controlled by potentiometer on switchbox. Model CBX-111. (Courtesy Mura Corp.)

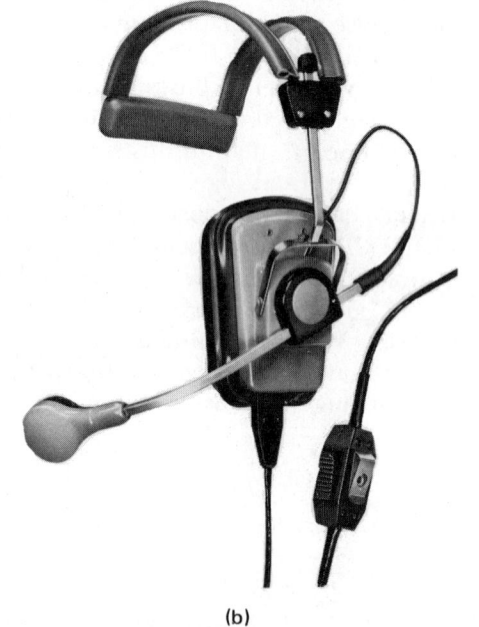

(a)   (b)

FIG. 8-8 Amplified boom microphone headsets. (a) Model CB-88. Lightweight set with noise cancelling magnetic microphone. (b) Model CB-1200 with boom mounted ceramic power microphone. (Courtesy Telex Communication, Inc.)

microphone is in a noise-canceling setup. Two versions of the CB headset–microphone combination made by Telex are shown in Fig. 8-8. The CB-1200 uses a ceramic microphone cartridge; the CB-88 features a noise-canceling dynamic transducer in the setup. Both Telex boom-microphone headset combinations have built-in amplifiers.

One variation of the headphone–microphone combination is the telephone handset, as shown in Fig. 8-9. A switch and switch button are placed at the center of the handset. Circuits are completed and broken by the switch as in the microphone in Fig. 8-1.

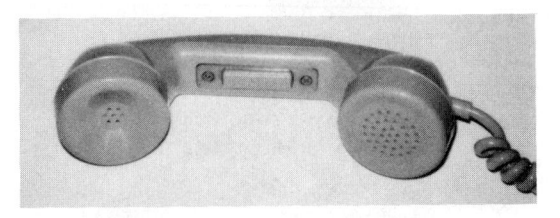

FIG. 8-9   Telephone handset with switch.

Amplified Microphones

Power microphones and their functions were described in Chapter 2. The Mura DX-116 with a slide-type gain control is shown in Fig. 2-9. The Mura DX-2000 base station microphone with a built-in amplifier is shown in Fig. 8-10. A microphone using a fixed gain amplifier, the Mura DX-120, looks very similar to that shown in Fig. 8-2. One big advantage of a microphone with fixed gain is that the level control cannot be reset accidentally to unwittingly overmodulate the transmission or provide so much gain as to put the entire transmitting system into an oscillating state.

The level control on a variable gain power microphone can be adjusted using a modulation meter such as that shown in Fig. 3-9. Connect the modulation meter to the antenna output jack on the transceiver. Do not connect the antenna to its jack on the meter, but instead use a 50-Ω resistor load. Set the level control on the microphone to minimum. Key the microphone on and adjust the controls on the tester so that you can read the percent of modulation. Whistle loudly into the microphone. Advance the level control on the microphone until the meter pointer just about indicates 100 percent modulation. Use this setting when transmitting.

If you do not have a modulation meter, communicate with another CB station, a "good buddy." Set the level control on the microphone to the point where you will produce the maximum distortion-free audio, as observed on the transceiver receiving your signal.

Two power microphones made by Turner are shown in Fig. 8-11. The microphone in Fig. 8-11a is primarily for mobile applications; the

FIG. 8-10 Base station power microphone with separate press-to-talk button and lock button for hands-free talking. Uses dynamic cartridge. Model DX-2000. (Courtesy Mura Corp.)

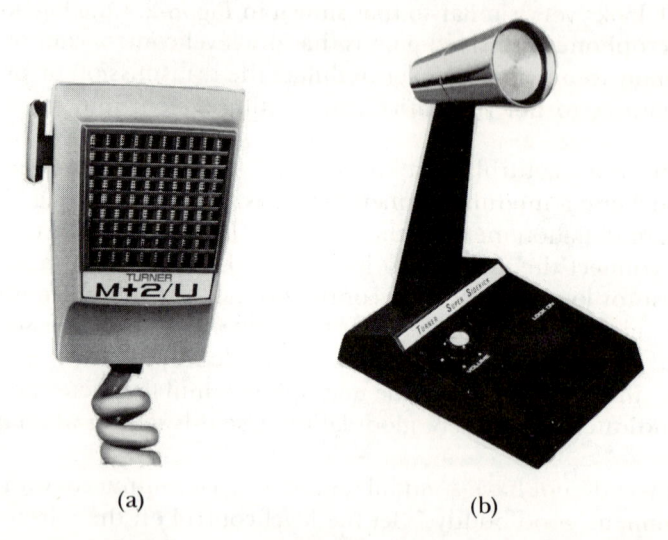

(a) (b)

FIG. 8-11 Power microphone. (a) Hand-held type using ceramic cartridge. Model M+2/U. (b) Base station type using dynamic cartridge-Model: Super Sidekick. (Courtesy Turner Division, Conrac Corp.)

Amplified Microphones

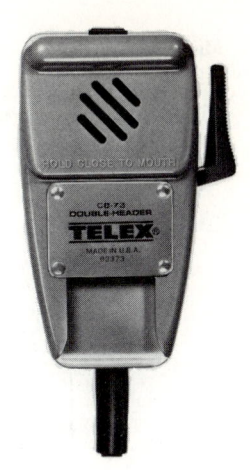

FIG. 8-12 Noise cancelling power microphone using dynamic cartridge. Model CB-73. (Courtesy Telex Communications, Inc.)

Super Sidekick in Fig. 8-11b has been designed for use with base station installations.

An amplifier along with the noise-canceling feature has been built into mobile microphone packages. The Telex CB-73 using this type of arrangement is shown in Fig. 8-12.

Installations using amplified microphones may be plagued by oscillation. It may manifest itself in the transmitted signal as hum, motorboating, or as a steady audio tone. This oscillation is not generated in the microphone. It can be traced to different aspects of the installation or to individual components in the setup. It is usually due to RF feedback from the antenna to the microphone.

The undesirable effects from this type of feedback can usually be eliminated by wiring a 10,000-Ω, 0.25-W resistor between the audio lead from the microphone and its terminal at the microphone connector. In severe cases, a 0.001 to 0.01 mF capacitor will have to be used in addition to the 10,000-Ω resistor. Connect the capacitor from the audio output terminal at the microphone plug to the ground or shield terminal. If output from the microphone (due to this modification) is not sufficient to drive the amplifier to 100 percent modulation, replace the resistor with a choke. Any value above 50 microhenrys (μH) should do the job.

Oscillation can also be eliminated by minimizing the SWR of the antenna system, grounding the transceiver properly to the auto chassis (or to a water pipe, radiator, etc., in base station installations), or just setting up a counterpoise. To do this, connect one end of an 8.5- or 9-ft insulated stranded wire to the transceiver chassis. Run the wire around the room or car. Do not connect the second end of the wire to anything.

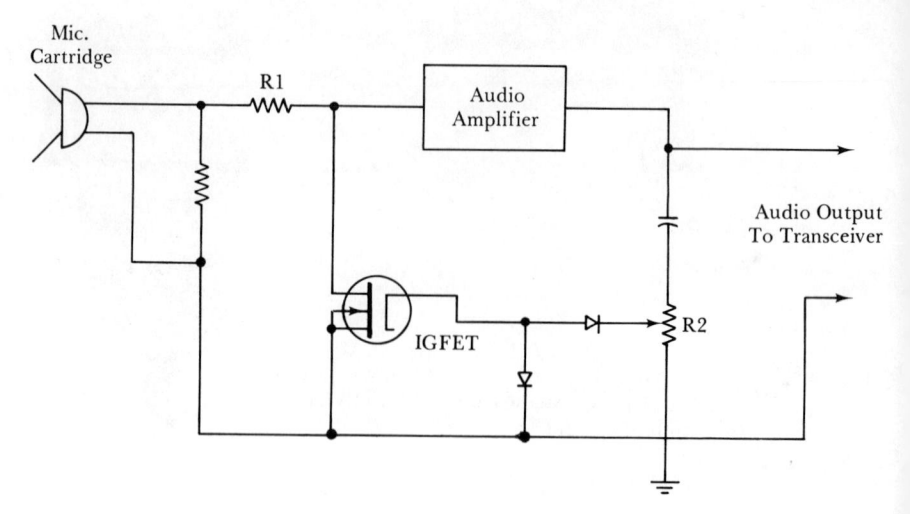

FIG. 8-13 Basic compressor circuit.

Compressor Amplifiers

As was stressed in earlier chapters, maximum power is delivered to the antenna when the RF is modulated 100 percent. Average power is at a maximum when this level of modulation is continuous. Speech consists of signal peaks. If the peaks of loud speech modulate the RF 100 percent, speech at ordinary levels modulates the RF at far less than 100 percent. To extend the time during which modulation is at a maximum, audio peaks must be limited or reduced when they exceed a predetermined amplitude. If large peaks are reduced to be similar in amplitude to average audio signal levels, excessive peaks would not exist. The size of average level audio signals can then be increased so that they modulate the RF near 100 percent for a longer period of time than would ordinarily be possible. There will be no problem of overmodulation by the peaks in the signal. Audio compressors are used to perform this function. A compressor may take the form of the modulation limiter built into every CB radio.

An FET functions well in a compressor circuit. IGFETs act as variable resistors. The resistance of the FET changes with the size of the negative voltage at the gate, decreasing radically as the voltage becomes more negative. A typical circuit that can be used in a microphone is shown in Fig. 8-13.

Signal from the cartridge is fed through R1 to the audio amplifier. Its output is applied to the transceiver as well as across potentiometer R2. A portion of the audio across the control is rectified by the diodes and fed to the gate of the IGFET. When the amplitude of the audio is

Peak Redistribution Modulation

low, the rectified negative voltage at the gate is small. The resistance of the transistor is large. Should this signal increase above a predetermined level, sufficient negative voltage is applied to the gate so as to reduce the resistance of the transistor. Acting as a voltage divider in conjunction with R1, the IGFET shunts some of the signal away from the input to the audio amplifier, reducing its output.

Should the audio from the cartridge increase, more negative direct current is fed to the gate, the resistance of the IGFET is reduced further, and a greater part of the signal from the cartridge never reaches the audio amplifier because it is shunted away by the very low resistance of the IGFET. This tends to keep the audio output at a constant level.

Two examples of compressor microphones are shown in Fig. 8-14, a Turner hand-held microphone and a microphone solely for base station applications.

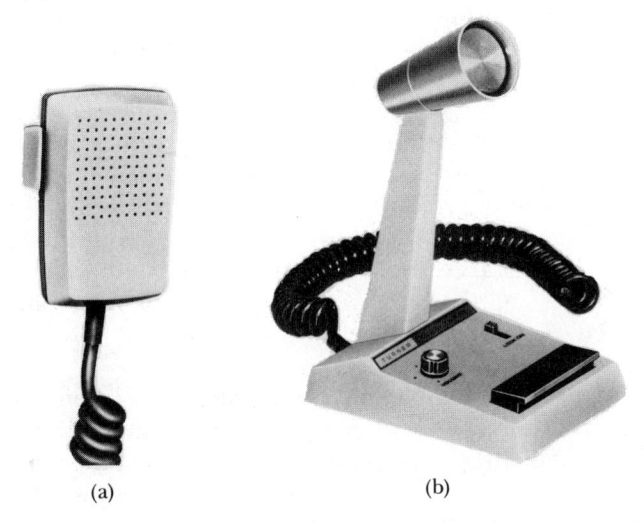

FIG. 8-14 Microphones using ceramic cartridges and incorporating compressor amplifiers. (a) Hand-held Model M+3. (b) Base station Model +3. (Courtesy Turner Division, Conrac Corp.)

Peak Redistribution Modulation

As was noted earlier, speech does not consist of pure sine waves, but is composed of peaks. They are frequently unidirectional. The entire peak is either above or below the zero axis. Some unidirectional peaks are so high that they tend to overmodulate the RF (force the RF excursion beyond its limiting positive or negative modulation maximum) more than would the same sized peak if it were divided into two equal portions, with one-half above and the other half below the zero axis.

Peaks are, of course, limited or clipped in the CB radio so that overmodulation is not likely. Distortion is generated, and a portion of the audio is lost because of peak limiting. It would be best if the peaks could be eliminated or made symmetrical around the zero axis, without adding to the distortion and without excluding a portion of the audio signal.

A new patented system referred to as peak redistribution modulation (PRM) has been developed at the Mura Corporation. It is designed to overcome the aforementioned drawbacks. The PRM circuit, complete with an audio amplifier for power microphone applications, is built into the microphone case.

Peaks generated by speech or any other signal source are composed of many sinusoidal-shaped frequencies, such as the one shown in Fig. 2-8b. The frequencies that combine or add to form the peak depend upon the amount of time that it actually exists, or the pulse width. The size or amplitude of each frequency component is also important in determining the various pulse dimensions.

The shape of the peak also depends upon the relative phases (relative time with respect to the lowest frequency at which each individual component cycle starts) of all frequency components. In the PRM system, the phases of the frequencies in the original signal are shifted with respect to each other, so that they no longer add up to the peak which generated them. Large peaks are minimized, and what remains are symmetrical around the zero axis. Their ability to overmodulate the RF is minimized. A typical signal before and after PRM processing is shown in Fig. 8-15.

The frequency components are amplified sufficiently to modulate the RF to 100 percent for a good portion of the cycle and deliver a maximum amount of power to the antenna. At the same time, the quantity and amplitude of audible harmonics present in the original speech signals have not been altered by a clipping or limiting circuit. Distortion of the reproduced audio through the loudspeaker is consequently reduced to a minimum.

Two examples of microphones using PRM circuitry are shown in Fig. 8-16. The Mura PRX-100 is for mobile and base station applications; the PRX-300 is designed exclusively for use with base stations.

The circuit and concept are patented. Four R–C phase shifting networks and two audio amplifier stages comprise the circuit, as shown in Fig. 8-17. Direct-current feedback is provided for stability. When troubleshooting, open the feedback loop and supply a fixed dc bias voltage to the transistor. To do this, connect a fixed dc voltage source through a 150,000-Ω resistor to the base of the input transistor. Then troubleshoot the circuit using ordinary voltage measurements, signal injection, and signal-tracing techniques.

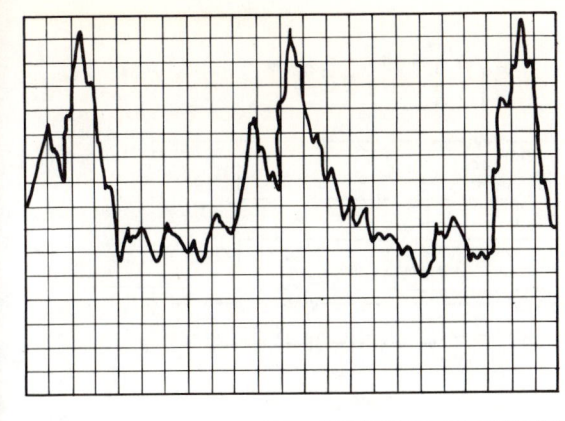

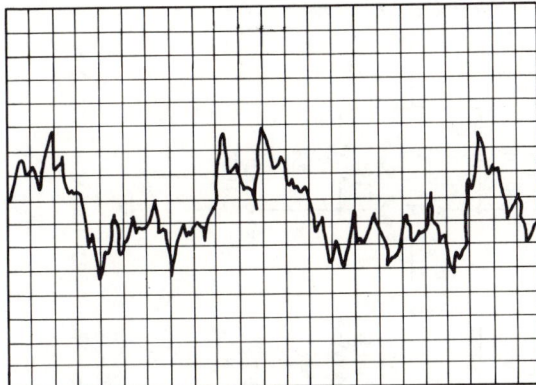

(a)
Input Signal Before Amplification:
Note Asymmetry and Sharp Peaks.

(b)
Same Signal After PRM Processing.

FIG. 8-15 A signal (a) before and (b) after being processed by PRM. (Courtesy Mura Corp.)

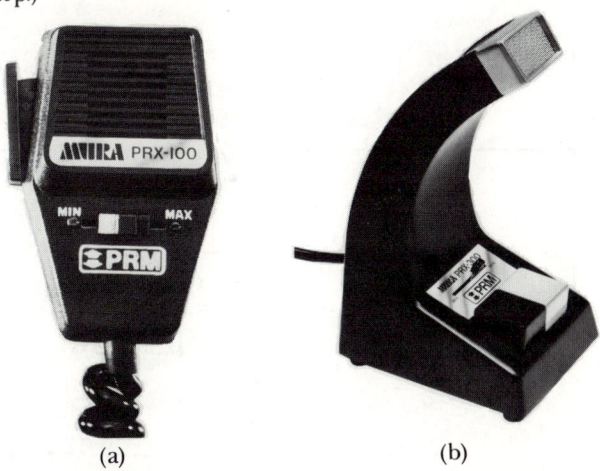

FIG. 8-16 Amplified microphones using dynamic cartridges and patented PRM circuitry. (a) Hand-held model PRX-100. (b) Base station model PRX-300. (Courtesy Mura Corp.)

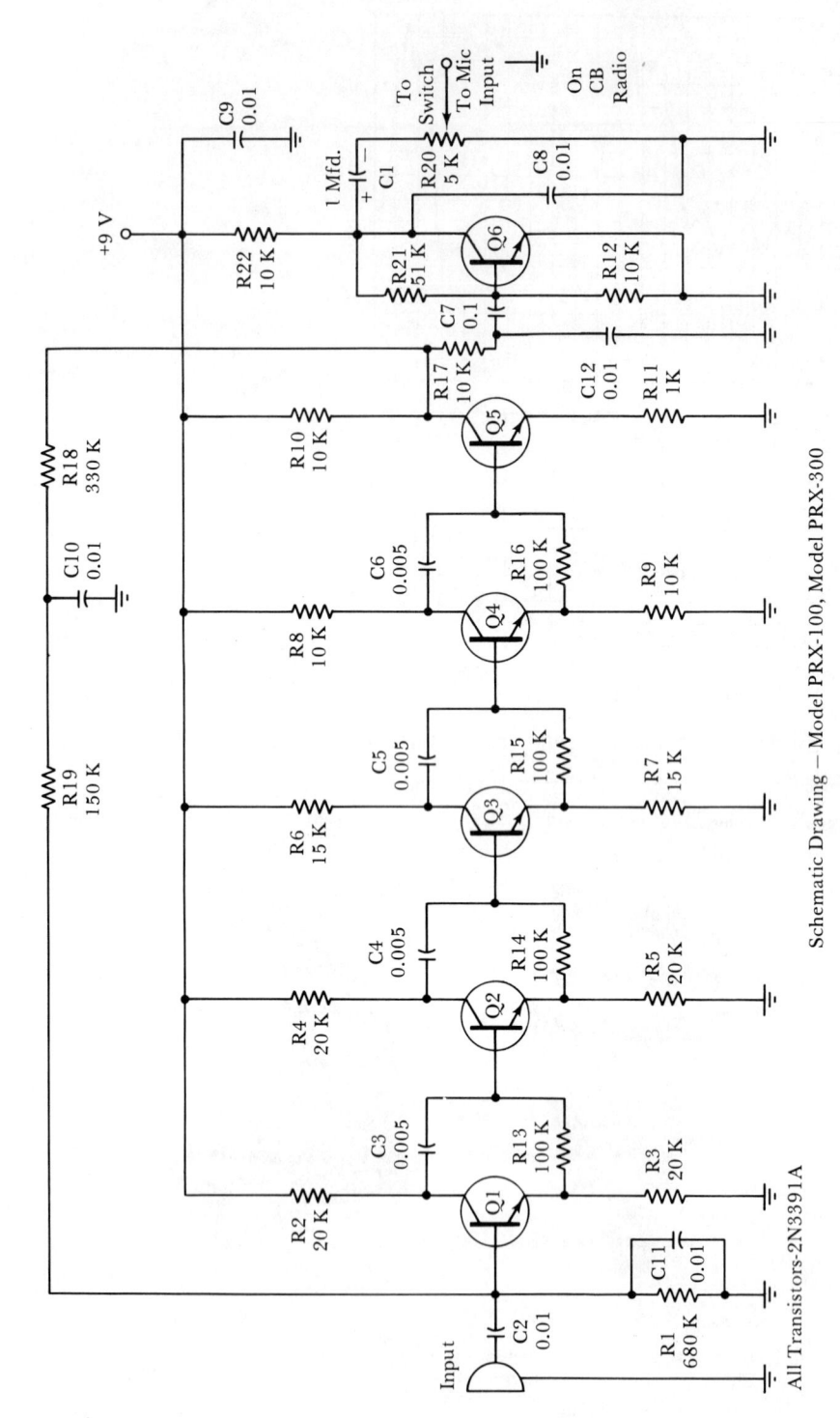

FIG. 8-17 Patented PRM circuit. (Courtesy Mura Corp.) Schematic Drawing — Model PRX-100, Model PRX-300

184

9

Antennas

The antenna is the component of the CB rig that has the maximum effect on transmission range or distance. Legally, no more than 4 W may be delivered by the transceiver. Maximum range can be assured only if the antenna is as efficient as it can be using modern technology. The height of the antenna has a bearing on efficiency, and height limitations have been set by the FCC. These limits primarily affect base station installations and have little or no meaning for mobile setups.

Roughly, an antenna system consists of up to four factors. There is, of course, the antenna proper. The antenna is useless unless it is connected to the transceiver, so we must consider the connecting coaxial cable as the second important component. All antennas must be mounted on some sort of structure. It can be an ordinary pipe on the roof or a pole or tower especially erected for this purpose. In mobile installations, the structure may be the roof of a car. Finally, some means is usually provided to rotate a directional base station antenna, so that there is no limit as to the direction in which transmission and reception are possible.

Antenna Characteristics

Antennas are complex items. Some basic characteristics should be defined before discussing specific installations, so that the pros and cons of the various setups and types of antennas can be appreciated better.

Feedpoint Impedance

Resistance is a measure of how well a conductor or circuit restrains the flow of dc current. It is the ratio of the voltage, E, across the circuit to the current, I, flowing through that circuit. Written mathematically, resistance is E/I. Alternating current is impeded by the presence of inductors and/or capacitors in the circuit. The restraining factor in low- and high-frequency (RF) ac circuits is referred to as *impedance*. The same E/I ratio that applies to resistance also applies to impedance. The only difference is the type of voltage applied to the circuit, whether it be alternating or direct current. In either case, impedance is measured in units called *ohms*.

Now let us turn to power. Power is the product of E and I. Ideally, 4-W RF power is delivered from the transceiver to a specified impedance. Current through and voltage across that impedance determine the power, $E \times I$, developed across it. Power from a CB radio should be fed to a 50-Ω antenna load. Thus, if E were 14.1 V, I would have to be 14.1 V divided by 50 Ω to yield 0.282 A ($I = E/R$) through the antenna, if 4 W is to be delivered to the 50-Ω load. To double-check this, multiply E and I; 14.1 $\times$ 0.282 = 4 W across 50 Ω.

The feedpoint impedance, or the impedance at the point where signal is fed to the antenna, is ideally 50 Ω. It is much higher at other points along the antenna rod. Store the ideal of 50-Ω input impedance or radiation resistance of the antenna in the back of your mind.

Short antennas exhibit much lower impedance at the feedpoint. It may be on the order of 5 or 10 Ω. Loading coils are used to increase these impedances to 50 Ω.

Half-Wave Dipole

Wavelength (the symbol for which is the Greek letter lower case lambda, λ) in feet of any RF signal is 984 divided by the frequency expressed in megahertz. As far as CB radio operation in the 27-MHz range is concerned, it is 984 divided by 27, or approximately 36 ft.

The length of one RF cycle traveling in free air is about 36 ft. When propagated through metal, such as the rod used as an antenna element, the speed of the RF signal is reduced. Because of the lower speed, the

Antenna Characteristics

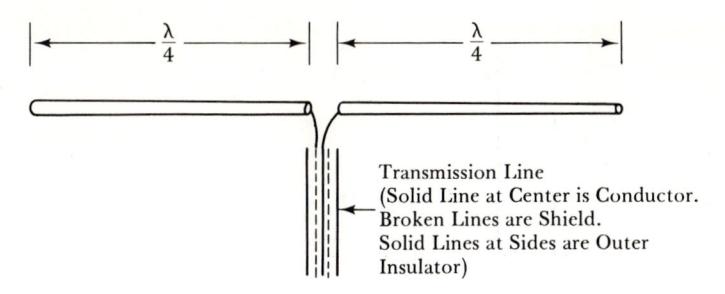

FIG. 9-1 Half wave dipole connected to transmission line.

length of the cycle is also reduced from the 36 ft just calculated. The 36 ft must be multiplied by a velocity factor of about 0.95, if the wavelength in the antenna rod is to be determined. One wavelength of a 27-MHz signal traveling in an antenna rod is about 0.95×36, or 34 ft.

Half-wave dipole antennas are a half-wavelength long, or are about 17 ft. These antennas are split into two sections, each being a quarter-wave long, or about 8.5 ft, as shown in Fig. 9-1. Each quarter-wavelength section is a hollow metal rod. If this horizontal antenna is oriented parallel to the ground, but a half-wavelength or more above ground level, its impedance at the connections to the cable is 72 Ω. Although the impedance of the antenna is much higher at its ends, only the feedpoint impedance at the center concerns us.

Most antennas used on CB rigs radiate and receive energy best in the directions in which they are oriented. Thus a horizontally oriented antenna at a receiver normally favors signals from a horizontal antenna such as the one shown in Fig. 9-1. The signal from this antenna is referred to as *horizontally polarized*. Most CB broadcasting is vertically polarized, so the dipole must be turned at 90° with respect to the ground, as shown in Fig. 9-2. Note that the coaxial cable is routed through the lower half of the hollow dipole rod so that it will not run alongside the antenna to disturb its radiation pattern. The impedance of this antenna is about 35 Ω.

A 50-Ω impedance is most desirable in CB applications as it has been accepted as the standard. This is a cross between the 72-Ω horizontally oriented antenna and the 35-Ω vertically oriented antenna. To achieve the 50-Ω goal, radials or rods are located near the bottom of (but not connected to) the ground plane antenna. The radials are usually connected to a grounded supporting mast. They are oriented at a 45° angle with respect to the ground, and thus droop. The radials form a ground plane for the antenna. Because the rods are neither horizontal nor vertical, but are at a 45° angle with respect to the radiator, the antenna impedance is neither 72 Ω nor 35 Ω. It is a cross (geometric

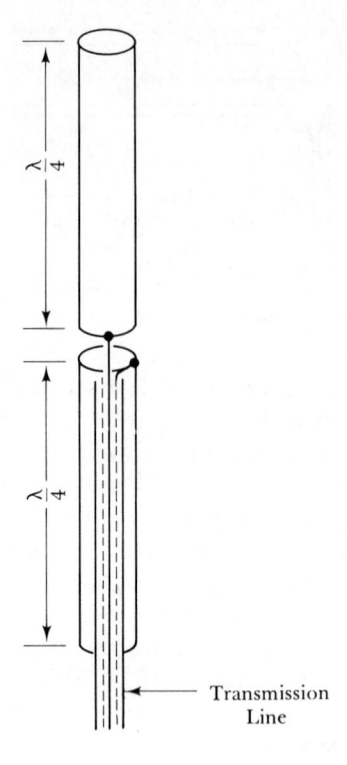

FIG. 9-2 Ground plane antenna.

mean) between the two. The half-wave dipole antenna now presents a 50-Ω load to the transceiver.

Directional Characteristics

As a general rule, signals radiated by horizontally oriented antennas are more directional than those from vertically oriented antennas. Maximum signal is generally radiated perpendicular to the rods that form the antenna. As for the half-wave dipole in Fig. 9-1, the maximum radiation is perpendicular to the page.

On the other hand, vertical antennas are omnidirectional. Signal is radiated in all directions. Objects near the antenna affect the radiation pattern. Trees and buildings at a base station installation, for example, affect the field.

Mobile antennas are in a separate group. The length of a half-wave dipole is 17 ft. It is composed of two 8.5-ft sections. Only one 8.5-ft section must be used on a mobile installation. If this 8.5-ft antenna is mounted on the roof of a car, an equal 8.5-ft length is reflected in the roof. The full half-wave dipole is thus formed by connecting the center

Antenna Characteristics   189

of the coaxial cable to the actual 8.5-ft length of rod and the shield to the metal roof. Located at the center of the roof, this antenna is nearly omnidirectional.

Should the antenna be mounted at one end of the roof, or on a bumper, a larger portion of the auto's surface is on one side of the antenna than on the other. Radiation is at a maximum in the direction of the largest surface of the auto's body (see Fig. 9-3).

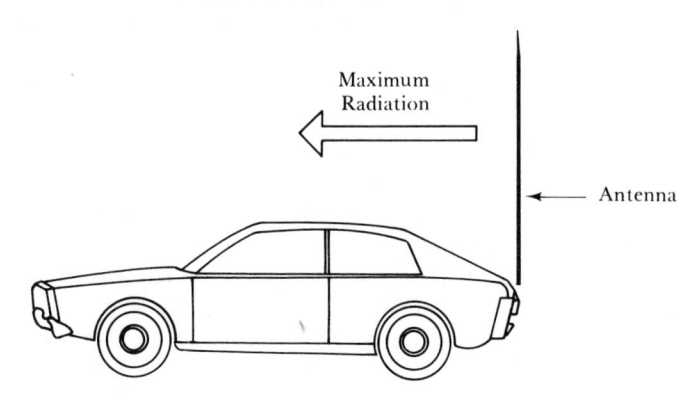

FIG. 9-3   Direction of maximum radiation from mobile antenna.

Radiation Angle

Signal is radiated at various angles with respect to the orientation of the antenna. Radiation from vertical antennas can have a zero-degree radiation angle with respect to the earth. In this case, the bulk of signal is parallel to the ground and is sometimes referred to as a *ground wave*.

Radiation angle is very dependent upon the design of the antenna. It is also dependent upon how high the antenna is mounted above the ground or ground plane. Generally, the radiation angle is smaller when the antenna is mounted at a higher elevation. Radiation angles of vertically oriented antennas are lower than the angles of those mounted in a horizontal plane.

When the radiation angle is high, signal is beamed toward the sky rather than parallel to the earth. Like light reflected from a mirror, radiation of the radio waves is reflected back to the earth at an angle equal to that of the original signal. (Like light, the angle of incidence is equal to the angle of reflection.) The reflected signal may reach the earth at a point several thousand miles away from the source. Because the FCC limits the range to 150 miles, this type of communication is illegal. It is also unreliable. Reflected signals may appear and disappear periodically and reception will be erratic.

Antenna Gain

The term "antenna gain" is somewhat misleading. Antennas do not amplify and do not have gain. Gain, as referred to here, is only a comparison of how much more apparent power a specific antenna radiates than does a standard antenna, when both are fed the same amount of actual power.

A theoretical isotropic antenna is referred to as the standard. This antenna is an imaginary point that radiates equally in all directions. Its gain is 0 dB.

More realistic radiating standards have been established, although they are also difficult to achieve physically. A horizontally placed half-wave dipole in free air is said to have power gain of +2.1 dB. Other antennas can be compared to this one.

It is simple to compare antenna performance through the use of Table 9-1. The isotropic antenna is a basic standard rated at 0 dB. Being the standard, its relative output when compared to itself is, of course, 1:1. The half-wave dipole is specified as having a gain of 2:1 dB. When compared to the standard, it appears to radiate 1.6 times more power than the standard. So if the standard can theoretically radiate 1 W when a specific signal is applied to it, the dipole will seem to radiate 1.6 W with the same input signal. This is an increase of 60 percent in efficiency. But it must be remembered that even the most efficient antenna cannot radiate more power than is applied to it from the transmitter. More efficient antennas can only radiate a bigger percentage of the applied power than can their less efficient counterparts. The gain figures show how much more of the signal will be radiated by one antenna than by another. An even more effective antenna may be specified as having a gain of 8 dB. All this means is that this higher-gain antenna will radiate 6.3 times the amount of power that the standard can radiate. The figure of 6.3 can be determined from Table 9-1.

Table 9-1 can also be used to compare two different real antennas. Suppose that an antenna is rated at 3-dB gain. The ratio of the power this antenna can radiate to the power radiated from the same signal by the isotropic standard is 2:1. Another antenna rated at 9 dB can radiate 8 times more power than the standard with the same applied signal. The power ratio here is thus 8:1. If you divide one power ratio by the other, the gain of one of these real antennas when compared to the other is 8:1/2:1 = 4:1. One antenna is four times more efficient than the other.

Another way of computing this is to subtract one decibel gain figure from the other. In our example, it is 9 dB less 3 dB, or 6 dB. Locating 6 dB on the chart, you find that the power ratio is 4:1, the same relative efficiency figure found by the preceding method.

Table 9-1   Relationship Between Decibels and Power Ratio

Decibels	Power Ratio	Decibels	Power Ratio
0	1:1	11	12.6:1
1	1.3:1	12	15.9:1
2	1.6:1	13	20:1
3	2:1	14	25.1:1
4	2.5:1	15	31.6:1
5	3.2:1	16	39.8:1
6	4:1	17	50.1:1
7	5:1	18	63:1
8	6.3:1	19	79.4:1
9	8:1	20	100:1
10	10:1	21	125.9:1

Supporting Structures

Antennas are not mounted in thin air and obviously require a base or structure to support them. The height of the overall antenna above ground is limited by FCC rules. It must conform to at least one of the following four criteria, or the installation is illegal. It does not have to conform in all ways.

1. The maximum height of the antenna must be less than 20 ft above ground.

2. The maximum height of the antenna must be less than 20 ft above any natural formation such as a tree, or less than 20 ft above a structure such as a house. A mast or pole of any sort is not considered a structure.

3. The maximum height of the antenna must be less than 60 ft above ground level if it is mounted on the antenna structure of another radio station. The highest point of the antenna mast must not extend above the height of the structure.

4. Any omnidirectional antenna must not extend to more than 60 ft above ground. Furthermore, if it is near an airport, find out the elevation figure of the nearest runway. The maximum height of the antenna must be less than 1 ft above this elevation for each 100 ft that the antenna is away from the runway. If, for example, the antenna is 300 ft away from the runway, the maximum permissible height above the elevation figure is 3 ft.

A location for mounting the antenna should not be chosen in a haphazard manner. The effect of an obstruction at the location depends upon the type of antenna used. Fields from omnidirectional ground plane or colinear antennas are influenced only by the proximity of other antennas or conductive objects. Quad or loop antennas, described below, must be located in a clear area at the maximum possible height above the ground. Beam or dual-phase types are not as dependent as are quad antennas upon how clear the surrounding area is, but do perform best under these ideal conditions.

The support structure should, of course, be strong enough to support the weight of the antenna used. Wind pressure should not be able to throw it over. Structures should be grounded and thereby serve as a return to earth for static charge and as a path for lightning to prevent the high voltage from reaching the transceiver.

If the load is such that the structure needs to be reinforced, guy wires should be used to anchor it firmly to the mounting surface. Guy wires can absorb energy from the antenna if the wires are resonant at frequencies in the 27-MHz range. To avoid this problem, place an insulator at about the center of each wire.

Rotators

Directional antennas are frequently rotated by motors controlled at the transceiver, so that the operator can choose the direction in which communication is optimized. Rotating equipment can be heavy, and the support structure must be strong enough to sustain the weight of the rotator along with the antenna.

To protect the rotator, the mounting should be such that the weight and tensions of the entire antenna assembly are on the support structure rather than on the rotator. This can be accomplished by mounting the mechanism several feet down inside the tower or mast.

The joint from the coaxial cable to the rotating antenna is important. Because the antenna may spin or "free-wheel" owing to wind, the proper attachment is important if the coaxial cable is not to be ripped off the antenna elements.

Coaxial Cable

Coaxial cable is just what its name implies. One conductor is inside a braided shield. They are separated by an insulating material. Another insulator is on the outside of the cable. The outer insulation protects the cable from the weather and environment.

An impedance exists between the two conductors. Impedance is not influenced by the length of the cable, but only by the spacing between the shield and the inner conductor. If the diameter of the inner conductor is represented by Di and the inner diameter of the shield by Do, the impedance of the cable, Zo, is

$$Zo = 138 \log Do/Di$$

Cables like RG-58/U, RG-58A/U, RG-8/U, RG-8A/U, and RG-213/U are dimensioned to have impedances of about 50 Ω. They are frequently used to conduct signal to the 50-Ω antenna.

Some cables are made with contaminating-type jackets. Life expectancy of these types is shorter than those with macks made of noncontaminating-type material. Of the five popular types mentioned, cables with noncontaminating-type jackets are RG-58A/U, RG-8A/U, and RG-213/U.

RG-8A/U and RG-213/U cables have a larger outer diameter than does the RG-58A/U type. The prime advantages of larger-diameter cables are added strength and the ability to handle high power. For example, RG-8A/U cable can handle about 2500 W at 27 MHz, whereas the RG-58A/U can handle nearly 400 W. Considering that the CB radio delivers only 4 W, power-handling ability is not a factor. Either type of cable will suffice.

Power is lost in the cable. In the 27-MHz range, losses due to each 100-ft length of RG-8A/U and RG-213/U cable are about 1 dB. The loss in an equal length of RG-58A/U cable is close to 2 dB. Shorter cables have proportionally less power losses. Whatever the loss, it must be subtracted from the antenna gain to determine the overall efficiency of the system. If long lengths of cable are needed in an installation, use the type exhibiting the lower losses.

One of the worst disasters can occur when a novice tries to connect the antenna plug, PL-259 (or Gold Line Model 228), to the end of the coaxial cable. If you follow the few simple steps outlined here, the job becomes an elementary procedure. The first group of steps is concerned with wiring a connector to the thick RG-8A/U and RG-213/U cables; the second group of steps covers the wiring procedures used for the thin RG-58A/U type.

Refer to Fig. 9-4a and proceed as follows:

1. Cut the RG-8A/U, RG-213/U, or any other like-diameter cable to the desired length.
2. Remove 1-1/8 in. of the outer jacket.
3. Clip off an 11/16-in. length of the braid shield so that only

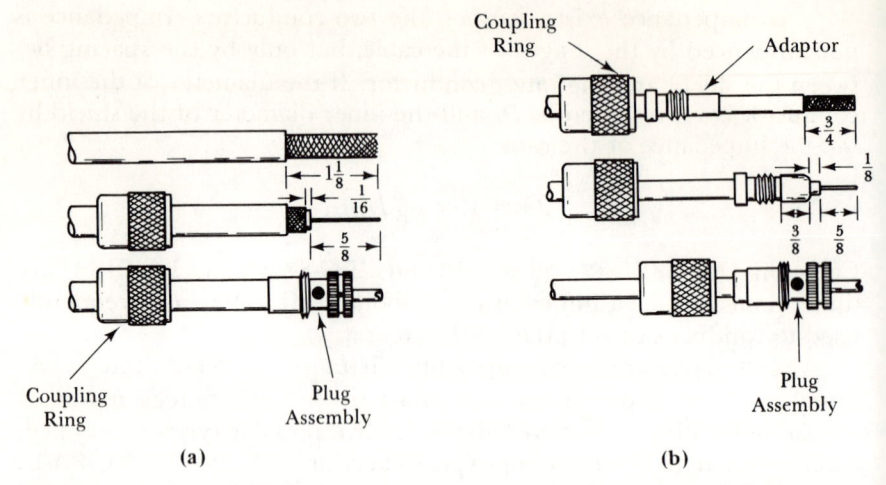

FIG. 9-4 Wiring of plug to transmission line. (a) To RG-8A/U and RG-213/U cables, (b) To RG-58A/U cables.

7/16 in. of shield protrudes outside the jacket.

4. Remove 5/8 in. of the inner insulator. A 1/16-in. length of the inner insulation must remain between the shield and the inner conductor so that they do not short to each other.
5. Tin the exposed portion of center conductor and the braid. Do not use excessive heat or you will melt the insulation.
6. Slip the coupling ring onto the wire with its threaded end toward the end of the wire just prepared.
7. Rotate the plug assembly onto the prepared end of the cable.
8. Make a good connection between the plug assembly and the braid by soldering one to the other through the holes. Do not use excessive heat so as not to melt the insulation. Be sure the solder is smooth and not grainy. Solder should not protrude above the outer surface of the assembly.
9. Solder the center conductor to the pin plug.
10. Slide the coupling ring onto the plug assembly.

If you use a thin cable such as the RG-58A/U, refer to Fig. 9-4b and proceed as follows:

1. Slip the coupling ring and adaptor onto the cable, as shown. The threads of the ring must be toward the end of the cable to be prepared here.

2. Cut this end of the cable to the desired length.
3. Remove 3/4 in. of the outer jacket.
4. Remove a 5/16-in. length of braid shield so that 7/16 in. of shield protrudes outside the jacket.
5. Spread the braid slightly and fold it back over the adaptor. Trim this shield so that only 3/8 in. of braid is over the adaptor.
6. Remove 5/8 in. of the inner insulation.
7. Tin the exposed portion of the center conductor. Do not use excessive heat or you will melt the insulation.
8. Screw the plug assembly onto the adaptor.
9. Complete the job using steps 8, 9, and 10 in the preceding list.

Water should be kept out of the coaxial cable by covering the exposed ends with a silicon rubber coating. By forming a drooping loop as shown in Fig. 9-5, you can keep water from entering a building along with the cable. Never let the diameter of any bend in the cable be less than 2-1/2 in. or a short may eventually be formed between the braid and inner conductor.

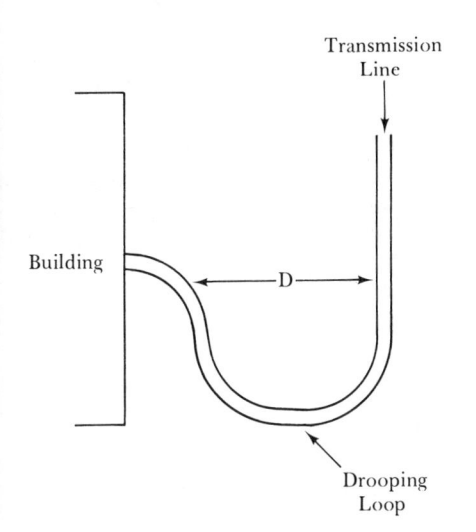

FIG. 9-5 Drooping loop to keep water out of building.

Base Station Antennas

The basic ground plane antenna consists of a vertical radiator and three or four horizontal radials. Each radial is a quarter-wavelength long. As discussed earlier, the impedance at the feedpoint is about 35 Ω. Impe-

FIG. 9-6 Ground plane base station antenna. (Courtesy Mura Corp.)

FIG. 9-7 Base station antenna with tuning ring. (Courtesy Cushcraft Corp.)

dance is increased to the desirable 50 Ω by letting the radials droop so that they are at an angle to the ground. Mura's version of this type of antenna is shown in Fig. 9-6.

Ground plane antennas are vertically polarized and omnidirectional. They radiate equally in all directions, depending of course upon the absence of an obstruction. If the radiator is a quarter-wavelength long, some of the power is radiated up to the atmosphere. The angle of radiation is about 50° with respect to the ground. The angle is reduced and made more parallel to the earth by increasing the size of the radiator. Radiators that are a half-wavelength in length radiate signals at about a 30° angle with respect to ground; by increasing the length to five-eights of a wavelength, the radiation angle is reduced to about 20°. Any increase beyond this does not substantially reduce the angle.

Base Station Antennas

Low-impedance 50-Ω cable can be connected directly to an antenna using a quarter-wavelength radiating rod. The cable is usually matched through a circuit when the radiator is one-half or five-eighths of a wavelength long. The Cushcraft CR1 antenna shown in Fig. 9-7 uses a large single-turn coil around the antenna. It is tapped at a point that will allow the antenna to present a 50-Ω impedance to the line. No radials are required here.

Directional antennas provide higher gain than omnidirectional types. Elements are added in front of and behind the signal rod to increase its gain and directivity. These antennas are all based on the yagi or beam type of antenna with the parasitic array so popular in television applications. The basic arrangement is shown in Fig. 9-8 with a reflector placed behind the dipole and a director in front of it. This arrangement adds to the power gain of the bare dipole. More elements can be added to the array. Each additional element increases the gain by about 1 dB. Tuning becomes difficult if more than six elements are used. This number of elements is the practical limit for CB rigs. One practical arrangement using five elements is shown in Fig. 9-9. Rotators are frequently used with this type of antenna to direct the beam in the desired direction. The antenna as shown is vertically polarized. Coincident horizontal and vertical polarization can be achieved when two such yagis are arranged at right angles to each other.

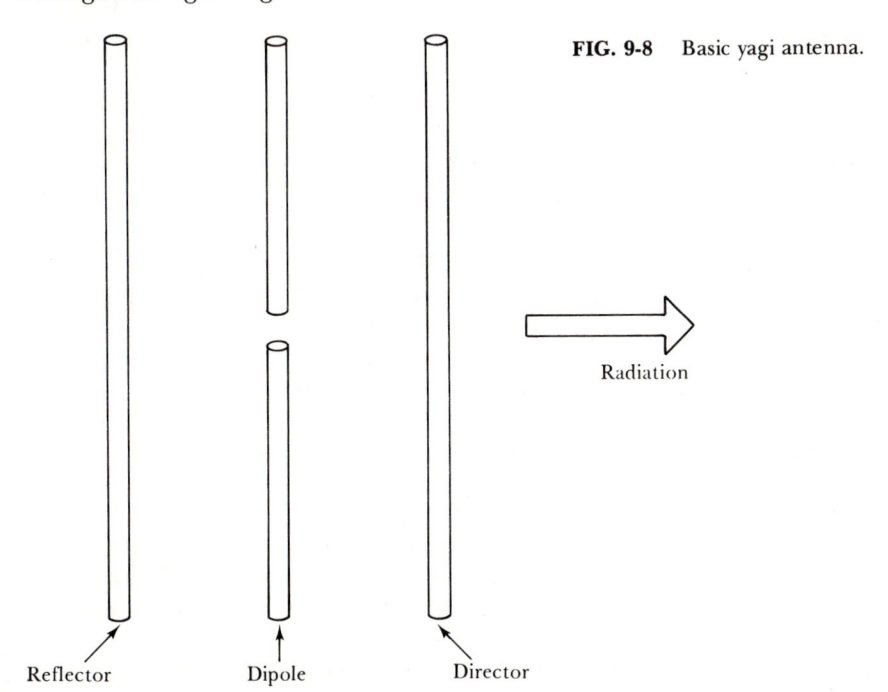

FIG. 9-8 Basic yagi antenna.

FIG. 9-9 Base station antenna using five-element array. (Courtesy Cushcraft Corp.)

Quad antennas are two half-wavelength metal rods arranged in the shape of a square. Each side of the square is a quarter-wavelength long so the total length of the metal rods adds to one full wavelength. Depending upon how it is mounted, it can radiate either vertically or horizontally polarized signals. Although its gain as a single element is quite good, it can be improved by adding square elements in front of and behind the actual radiator, similar to those associated with the simple yagi.

Phased arrays consist of two or more vertical antenna rods spaced a quarter- or half-wavelength apart. Because cable slows down the flow of RF, proper lengths of cable must be connected to the two antennas to achieve the desired radiation pattern. If the space between two antenna rods is a half-wavelength and signals are fed in phase to both antennas (assuming that the lengths of the cables connected to both are equal), narrow beams are radiated perpendicular to the two antennas. This is shown in Fig. 9-10a. The pattern changes to that in Fig. 9-10b if one lead from the cable is a half-wavelength longer than the other.

If antenna rods are spaced a quarter-wavelength apart, and the length of the cable to one antenna is a quarter-wavelength longer than the cable to the second, a radiation pattern similar to that shown in Fig. 9-11 is generated. It is highly directional. This directional characteristic is utilized to advantage in three antenna arrays where each radiating rod is located at one corner of an equilateral triangle. The length of each leg of the triangle is equal to a quarter-wavelength. The radiation pattern is selected through switching rather than through the use of a mechanical rotor.

A switch is located at the transceiver. Two of the radiating rods as

Base Station Antennas

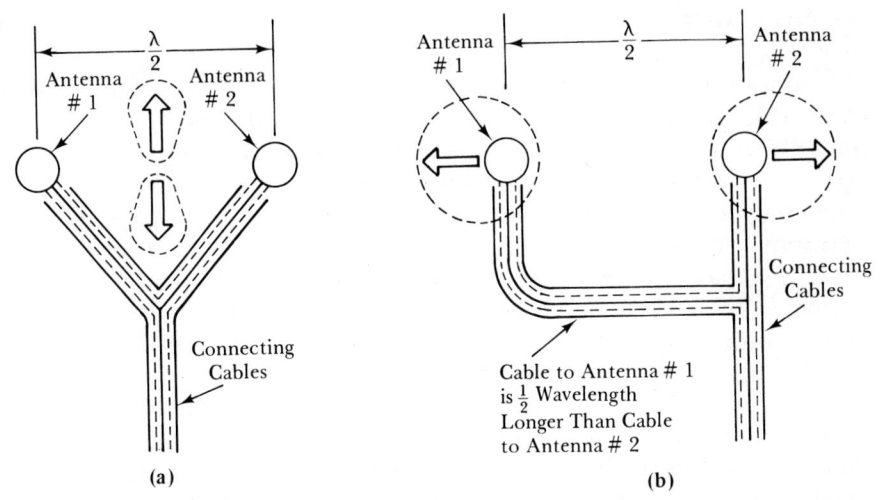

FIG. 9-10 Phased arrays using antennas that are one-half wavelength apart. Note radiation pattern from arrows and broken lines enclosing arrows. (a) Connecting cables to both antennas are equal in length, (b) Connecting cables to antenna no. 1 is one-half wavelength longer than to antenna no. 2.

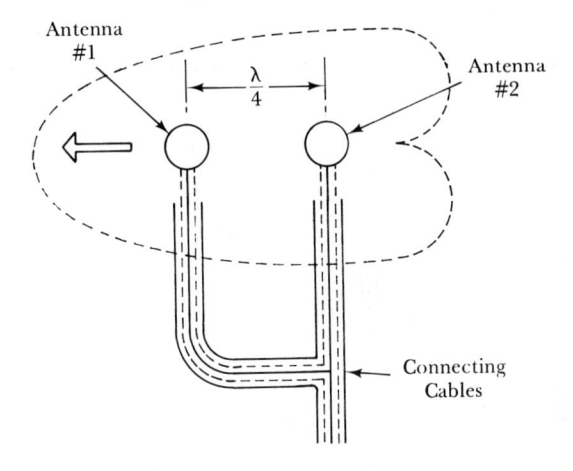

FIG. 9-11 Phased array using antennas that are one-fourth wavelength apart. Note radiation pattern from arrow and broken line enclosing arrow. Connecting cable to antenna no. 1 is one-fourth wavelength longer than to antenna no. 2.

well as the length of cable to each of the rods, are selected through the switch. Radiation patterns are controlled by altering the selection of rods to be used as radiators as well as the relative phases of applied RF to the respective antennas.

Mobile Antennas

Antennas for cars and trucks are available in different shapes and sizes. As a rule of thumb, longer antennas are more desirable, so use the longest antenna practical in your installation. If you consider the quarter-wavelength or 8.5-ft antenna as 100 percent efficient, a 6-ft antenna using a loading coil at its base (base loaded) is only 80 percent efficient (see Fig. 9-13). This is not too bad. From there on, the efficiency of a base-loaded antenna drops rapidly with respect to size. A 4-ft antenna is only 40 percent efficient, a 2-ft antenna is 20 percent efficient, and so on down the line to even smaller radiating devices. Efficiency is much higher when the loading coil is at the center of the antenna (center loaded), as in Fig. 9-12, but many good antennas are base loaded. An advantage of this type of antenna is that it is not top heavy and does not bend or break as readily as a center-loaded antenna would in the wind or while the vehicle is in motion. An efficient base-loaded antenna is shown in Fig. 9-13.

    Loading coils affect the efficiency of the antenna, but are required when it is shorter than a half-wavelength. Loading coils made with heavy wire have what is referred to as *high Q*. They absorb little energy by themselves, letting more of the energy from the transmitter reach the

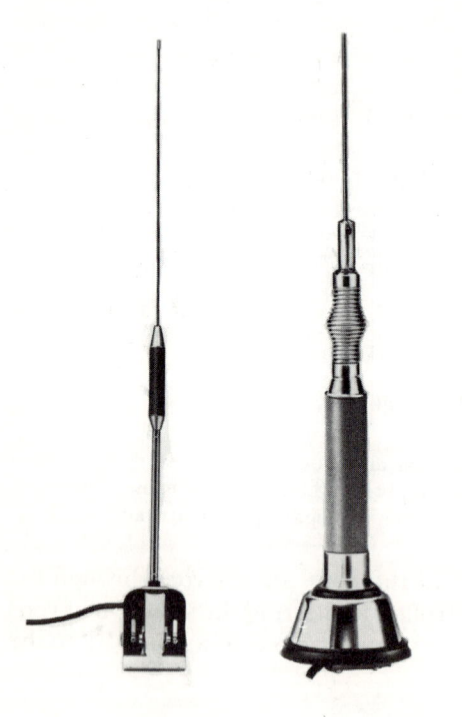

FIG. 9-12   (*Far left*) Center loaded antenna for mounting on rain gutter of car. (Courtesy Mura Corp.)

FIG. 9-13   (*Left*) Base loaded antenna for mounting at trunk lip or at center of trunk lid or roof. (Courtesy Mura Corp.)

Mobile Antennas

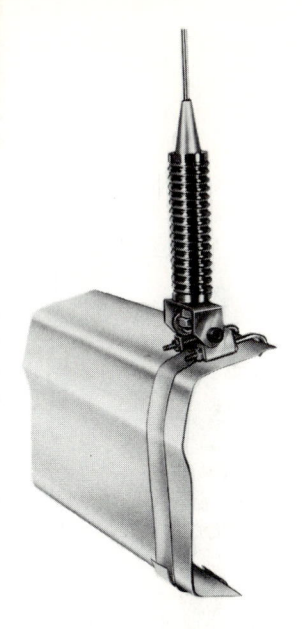

FIG. 9-14 Clamp for mounting antenna on bumper. (Courtesy Turner Division, Conrac Corp.)

whip for radiation. Low-Q coils reduce the quoted efficiency figures even further.

Antennas are mounted in different ways and at different locations on a vehicle. The best arrangement is a quarter-wavelength antenna mounted at the center of a roof. It is not usually practical to do this as it is 8.5 ft in length and may therefore hit the bottom of an underpass on a highway, a branch of a tree, or the like. Quarter-wavelength antennas are best mounted on the bumpers of cars using clamps, such as shown in Fig. 9-14. Somewhat shorter antennas, such as shown in Fig. 9-13, can be mounted at the center of the trunk or roof. Mounting at the center of an extended surface necessitates drilling a hole in the car. This affects its resale value. This drawback does not occur when one of the antennas in Figs. 9-12, 9-13, or 9-15 is used. The antenna in Fig. 9-12 can be mounted in the rain gutter of the vehicle; those in Figs. 9-13 and 9-15 can be mounted on the trunk lip. Neither type requires drilling holes.

Cophased mobile antennas have been discussed in Chapter 2. They consist of two radiating elements that should be mounted about 9 ft apart. Compromises are made when this spacing is not possible on a particular installation. A practical compromise involves two antennas for mounting in rain gutters on both sides of the car, as shown in Fig. 9-16. Parts to construct a larger cophased antenna are shown in Fig. 9-17. Clamps are supplied for attaching each antenna to a rear-view mirror at the two sides of the truck cab. Note these clamps at the center of the figure.

FIG. 9-15 Bracket for mounting antenna at trunk lip. (Courtesy Turner Division, Conrac Corp.)

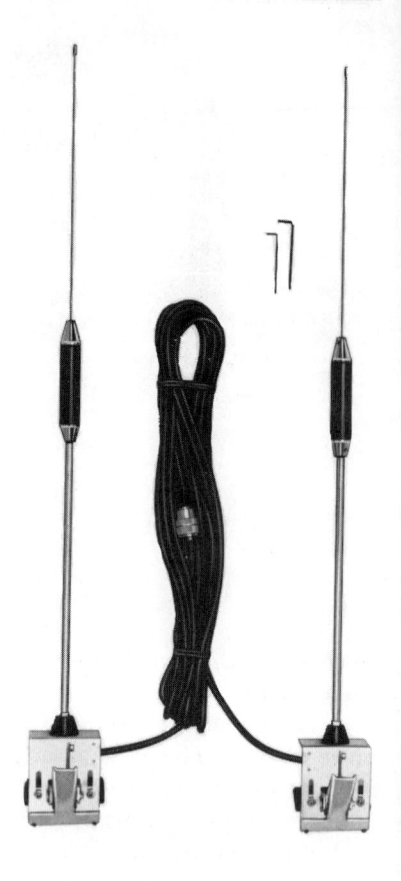

FIG. 9-16 Cophased antenna system to be mounted on rain gutters of car. (Courtesy Mura Corp.)

An antenna produced by Bill Owen, Inc., looks very much like a luggage rack on top of a car or station wagon. It is useful where height is a limiting factor, such as on a mobile home. The antenna height is less than 1 ft above the roof of the vehicle. Although the antenna is horizontal, radiation is said to be vertically polarized. Gain, SWR, and frequency range are claimed to be excellent. The design is protected by a U.S. patent issued to William G. Owen on February 4, 1975.

A protruding antenna is a dead give-away that a CB radio is in the vehicle. Hide the antenna and you hide the existence of a transceiver. One procedure is to use a magnet-mount antenna, such as the one in Fig. 9-18. The antenna is held to the roof or trunk lid by a powerful magnet in its base. It can be removed and stored in the trunk when the radio is not in use.

Tenna Corporation has another solution. They produce an antenna that can be extended to its full length only when the CB radio is in use. It is retracted at all other times (see Fig. 9-19). All this can be

Mobile Antennas

FIG. 9-17 (*Left*) Cophased antenna system for mounting on mirror brackets on trunk. Note clamps at center of photo. (Courtesy Mura Corp.)

FIG. 9-18 (*Below*) Mobile magnet mount antenna. (Courtesy Turner Division, Conrac Corp.)

accomplished with a simple flip of the switch at the transceiver. Convenience can be enhanced by placing the antenna in a circuit where the transceiver is turned on when the antenna is extended and turned off when the antenna is retracted. Being center loaded, it is quite efficient.

The International CB Radio Operators Association is now (1977) selling a ground plane antenna to be installed inside the motor vehicle. They refer to it by the trade name "Invisible Ear." It has facilities for tuning the installation and a tuning light indicator to show when this operation has been properly accomplished. It can be installed in a fashion so as to make it into a truly unobtrusive device. No gain or efficiency data has been supplied by the vendor. You should secure this information before proceeding with purchasing and installing this apparently desirable device.

Amplifiers have been designed to increase the gain of the antenna or RF signal when receiving. The Solar Hot Rod in Fig. 9-20 does this and more. It is an amplified antenna designed to be attached to your

FIG. 9-19 Antenna can be extended or made to drop out of sight by merely flicking a switch. (Courtesy Tenna Corp.)

FIG. 9-20 Solar powered—solar hot rod—CB antenna system. (Courtesy Lynn-Paul Marketing, Inc.)

present CB antenna. Its presence has no effect on the SWR. The active devices in the system are supplied energy from power stored in the Hot Rod. This power is, in turn, derived from solar sources. Received signals are increased up to 4 dB in amplitude, thereby improving the signal to noise ratio.

Whatever antenna is chosen, good installation is a very important factor in its proper performance. The antenna must be secured firmly to its support. Be sure to ground the mounting bracket to the support structure, if the installation instructions tell you to do this. If you use a gutter, trunk lip, or roof antenna, scrape off enough paint from the vehicle to be certain that the grounding is proper.

Tuning the Antenna

Most antennas are tuned for minimum SWR by adjusting the length of the whip. This adjustment is especially critical when a short antenna is used in an installation. An additional drawback when using a short antenna is that SWR is not uniform over the entire CB band. Should the whip length be adjusted for a minimum SWR when using signal from a channel at the center of the band in the adjustment procedure, SWR may be as high as 3:1 on channels at the extreme frequencies. As you are aware from Table 3-1, 11 percent of the power never reaches the antenna if the SWR is 2:1; 25 percent is lost if it is 3:1. It is thus very desirable to keep SWR to a maximum of 2:1 on all channels, but never more than 3:1.

If you are to use but one channel in the CB band, adjustment is simple. Set the transceiver to this channel. Adjust the whip length through trial and error for a minimum SWR. Use one of the meters described in Chapter 4. Move the whip up and down 1/8 in. at a time in its mount until you find the point where SWR is at a minimum. In general, you know to shorten the length of whip sticking out of the antenna if SWR increases as your hand is placed near it and to increase the length of the exposed portion of the whip if your hand capacity causes the SWR to decrease.

Two practical limitations must be considered when adjusting SWR: (1) inexpensive SWR meters are not accurate, and (2) SWR on some antennas cannot be adjusted to anywhere near 1:1 or even 2:1. Neither item should faze you. Adjust the whip length for a minimum SWR and accept this as the best you can do with the antenna involved.

If you find that you cannot adjust the whip to provide a minimum SWR reading, it may be necessary to cut the length of the whip or to obtain a longer one. If SWR is at a minimum only when the radio is set to channel 1, the whip is too long; if it is at a minimum only when using channel 23 or 40 (depending upon the CB radio), the whip is too short. SWR should be adjustable to a minimum on the channel that you want to use. If this cannot be accomplished, the length of the whip must be altered. Most of the time, the whip length as supplied by the manufacturer is satisfactory for all adjustments.

You cannot place the entire blame for high SWR on the antenna proper. SWR can frequently be reduced by moving the antenna to another point on the vehicle, or by moving the vehicle to another location away from another structure. There is always one optimum location for the antenna on the car, such as at the center of the roof and at a point far removed from other vertical rods. SWR can also be affected by weather conditions.

If the transceiver is to be used over the entire CB band rather than on only one channel as discussed, optimum adjustment criteria become more critical. SWR must be minimized at the center of the band accommodated by the transceiver involved. If it is a 23-channel radio, use channel 12 when adjusting SWR. Use channel 19 on 40-channel rigs. Once adjusted on the "center channel," check SWR at extreme ends of the band. Note SWR on channels 1 and 23 on 23-channel transceivers and on channels 1 and 40 on 40-channel CB radios. SWR should be less than 2:1 when using 3.5-ft or longer antennas and less than 3:1 or 4:1 on 2-ft or shorter antennas. As a larger band of frequencies is covered on 40-channel than on 23-channel transceivers, avoid short antennas on rigs for 40 channels. Short antennas are satisfactory for use with 23-channel radios.

Dual-phased antenna systems require more work when they are being adjusted for minimum SWR. You will have to go back and forth between the two antennas, adjusting both until settings are found where SWR is minimized.

If SWR cannot be minimized on cophased or single-whip antennas, matchers or tuners are sometimes recommended. They are placed between the transceiver and connecting cable (see Chapter 2). Although they do little to change the impedance of the antenna, they provide more accurate coupling of the transceiver to the coaxial cable. Output is improved when the matcher is adjusted for a minimum SWR, although in reality there are still standing waves on the line. Therefore, it is wise to adjust the antenna for a minimum SWR without this box and then insert the box between the SWR meter and line to improve the situation. An instrument combining an SWR meter and matcher is shown in Fig. 9-21.

Another method of achieving a low SWR is to cut the coaxial cable until you find the optimum length that will give you a minimum SWR reading. If changing the cable length changes the SWR reading, you are already in trouble. You can cut the length to give a minimum reading,

Tuning the Antenna

FIG. 9-21 Combination SWR and power meter with antenna matcher. (Courtesy Mura Corp.)

but you will still have enough reflected signal to make your measurements meaningless. Cutting the cable to an optimum length does not improve the actual SWR of the antenna system. Only the antenna should be adjusted for minimum SWR, not the transceiver or the line.

Although optimum efficiency may not be achieved through use of an antenna matcher or by cutting the coaxial cable to the length providing a minimum SWR reading, there is one tremendous advantage gained through either procedure. Transmission from some transceivers is erratic unless their output circuits are 50 Ω. The output transistor may even break down. A matcher or properly cut cable can help establish the desirable 50-Ω loading condition.

Once all adjustments have been completed, the range of the transceiver has been optimimized. Good base station installations can provide a range of up to 50 miles, although 20 miles is the more likely limit. Mobile stations transmit up to 5 miles over flat terrain; under unusually optimum circumstances, communication is possible at distances up to 10 miles.

10
Problems Encountered in the Field

Regardless of how well a transceiver may perform in the shop, some defects become obvious only after the CB radio has been installed in a vehicle. Weaknesses in the rig may or may not be related to the set itself. As an example of a field problem, consider the possibility that the CB radio is mounted in a stream of hot air. Transistors may be damaged and frequencies can shift owing to high temperature. In this case, all you can do is bring the set into the service shop, replace any damaged devices, and remount the radio at a cooler location in the vehicle.

Components may also be damaged when exposed to extremely cold temperatures. While transistors can ordinarily withstand temperatures somewhat below $-55°$, crystals cannot. If a CB radio is exposed to very low temperatures, suspect the crystals before considering other components as the cause of any defects. Of course, a dead radio may also be due to a dead battery when it is exposed to low temperatures. The radio should be checked in the service shop after it has been exposed to normal room temperatures for an hour or so.

Other field problems may be due to items other than the transceiver, such as the power supply, microphone, and antenna. Besides problems that may arise with the part proper, it must be installed properly using proven techniques if the overall procedure is to be financially lucrative.

Noise due to various components in the vehicle may appear when receiving signals. Automatic noise limiters are usually not sufficiently effective in keeping all objectionable noise out of the reproduced audio. You must identify the source of noise in the vehicle and minimize its effect. This is an important and time-consuming factor that must be dealt with after the basic installation has been completed.

Installing the CB Radio

The first decision concerning the installation must come from the user, who must decide if it is to be a permanent or temporary setup. Some people want to remove the antenna and CB radio from the car and store them in the trunk as a safety measure when the rig is not being used; others may switch cars from time to time or use a rented vehicle. A permanent installation is definitely inconsistent with their requirements.

Little effort need be extended when making a temporary installation; the completion of a permanent setup can require well over an hour of exhaustive effort.

Temporary Installations

A characteristic of every temporary installation is the built-in flexibility for easy removal of the transceiver and antenna from the vehicle. Magnet-mount antennas are recommended in these applications. Place the antenna on the roof of the car, run the cable through a barely opened rear window, and connect it to the transceiver. Adjust the length of the whip for a minimum SWR.

You can stir the wrath of a customer by placing an antenna on the bare roof of the vehicle. Although the base of the magnet-mount antenna is usually well padded so that it should not scratch the paint, scrapes or lines in the car's finish are inevitable. Avoid confrontations. Place a thin cardboard or padded fabric under the base of the antenna. Instruct the customer to lift the antenna up off the roof and not to slide it.

The transceiver can be placed on the seat next to the driver. Connect the antenna and microphone to the transceiver and the power leads, through a plug, to the cigar lighter socket. The transceiver is now operable.

Installing the CB Radio

Many temporary installations use a slide-in bracket assembly constructed of two basic parts; one part is permanently installed under the dashboard and wired to the antenna and power supply. The second is mounted onto the CB radio and connected to it. Slide the bracket on the transceiver into the one mounted in the vehicle, and the transceiver is ready for use. All interconnections have been completed through contacts on the brackets. Slide the radio off the permanently mounted bracket when you want to store it. If you also want to store the antenna, you must, of course, disconnect it from the permanently installed portion of the bracket.

Permanent Installations

The operator must make a choice as to where he wants to locate the CB radio. Before settling on the spot, attach the mounting bracket (supplied with the transceiver) loosely to the radio. Place it in the selected location to determine if it will fit into the preferred area. If it does, let the operator check if the location is convenient. He must decide for himself if he can readily see the panel and manipulate all controls. The set should not be so far from the operator that he must put excessive strain on the microphone cord each time he transmits.

Next remove the mounting bracket from the transceiver. If the transceiver is to be mounted under the dashboard, place the bracket against the appropriate spot on the dashboard. Use the bracket as a template to locate and drill the mounting holes. Secure the mounting bracket to the dashboard and the transceiver to the bracket.

If the transmission hump is to be used for mounting the transceiver, the bracket in Fig. 10-1 is a convenient device. Using the bracket

FIG. 10-1 Bracket for mounting transceiver to transmission hump in car. (Courtesy Mura Corp.)

as a template, locate the holes to be used for mounting the bracket onto the transmission hump. Secure the bracket to the hump, fasten the mounting bracket supplied with the transceiver to the hump mounting bracket, and then secure the transceiver to its bracket.

Next select an appropriate antenna using the information given in Chapter 9. Select a mounting location and fasten the antenna securely to that spot. Be sure that all necessary grounding is properly executed. Run the cable through the car, under rugs and seats and along the sides, so that it does not interfere with the operation of the vehicle or with passengers. If you locate the antenna at the trunk, it may be necessary to remove the rear seat and snake the antenna wire from the trunk into the passenger compartment through an opening. Fasten the connector at the end of the coaxial cable securely to the antenna jack on the transceiver.

Finally, connect the positive and negative leads from the transceiver to a source of dc power. The negative lead from the transceiver is normally connected to the vehicle's chassis ground or negative (−) terminal of the battery; the positive lead may be connected either to the accessory terminal of the ignition switch or to the hot lead of the cigar lighter, or to the positive (+) terminal on the battery.

A microphone hanger is usually installed at some convenient location to hold the microphone when it is not in use. The location should be chosen so that the microphone is conveniently accessible when the operator decides to transmit.

Now use the various procedures and instruments discussed in previous chapters to optimize the efficiency of the antenna system.

Noise and Interference

Noise originating in the electrical system of a vehicle can interfere with any CB installation. Its origin can be traced to the generator, spark plugs, distributor, ignition coil, or to any less obvious item where an undesirable spark gap may develop or a poor connection may evolve. Any electrical arcing is undesirable as it will transmit noise.

Before adding any gadgets to eliminate noise, be sure that everything now in the vehicle is in proper shape. Check all electrical connections to be sure that they are tight. Clean these connections and all wiring. Have a general tune-up, paying special attention to the spark-plug gaps. Any crimped wires to the plugs and distributor should be soldered to prevent a gap from forming between the wire and connector. Finally, check generator and alternator brushes and rings. They must be clean and in good shape.

One point frequently overlooked is the exhaust system. It, as well as

Installing the CB Radio

the engine, is generally insulated from the frame of the car. In this position, the exhaust pipe can act as a CB antenna radiating noise into your CB radio. It is wise to connect the pipe to the frame of the vehicle with braid straps.

Sources of noise can usually be identified by their effect on reception. Although one source can predominate, once it is eliminated another cause of interference will become evident. The procedure is to first try to eliminate the most annoying source of noise and then to minimize the remaining factors in turn, as they manifest themselves.

Popping noises are most prevelant. If the frequency of the pops increases with engine speed, the source of the interference is most likely in the ignition system. Check the plugs, distributor, and the wiring to these components. Noise from these items can frequently be minimized by connecting a 10,000-Ω resistor between the ignition coil and the distributor cap and a 5000-Ω resistor at each spark plug. The resistors should be rated to withstand any instantaneous voltages that may be impressed across them.

A good alternative is to use resistor spark plugs as well as shields over the plugs, distributor, and ignition coil. Ignition cables should be checked for breaks in the insulation. If they must be replaced, it is wise to use shielded cables throughout especially made for this purpose.

If the problem is due to the ignition coil, mount a feed-through or coaxial capacitor to a good ground on or close to the body of the coil. Break the dc path to the coil by disconnecting the wire from the coil to the ignition switch or battery terminal. Connect the wire to one lead from the center plate in the capacitor and the free terminal on the coil to the remaining lead from the center plate. The dc path is restored to the coil through the leads at the two ends of the center plate, while the outer shell of the capacitor acts in conjunction with the center plate to bypass noise pulses to ground. A 0.05- to 0.1-mF, 400-V capacitor such as the one in Fig. 10-2 can be used.

Popping noises can also be due to static charges formed by the spinning of the car's wheels or tires. In the former case, a static suppressor spring may be inserted between the axle and each front wheel. Should the tires be the cause of the noise, carbon in powder form may be fed into the tire. Both items are available off the shelf at good stores specializing in auto parts.

If your car is equipped with a voltage generator rather than an alternator, the following applies. If there is *clicking only while you are accelerating* and it stops when the engine idles, the relay contacts for the field coil of the voltage regulator may be at fault. There are several procedures that may be pursued to try to eliminate this source of noise. One possible solution is to connect a small 0.02- to 0.05-mF, 50-V

FIG. 10-2 Feed through capacitor. (Courtesy Gold Line Connector, Inc.)

capacitor in series with a 3- to 5-Ω resistor, and place this combination between the field (F) relay terminal on the regulator and a nearby ground. The broadly tuned filter in Fig. 10-3 may be used for this purpose.

In bad cases, use a 0.1- or 0.2-mF feed-through capacitor in series with the battery (B), ignition (I), and armature (A) terminals of the regulator box. It is also helpful to connect the center conductor of a piece of coaxial cable between the field terminal on the alternator and the field (F) terminal on the regulator, rather than use the ordinary wire originally supplied. All outer terminals of the capacitor and the shield of the cable should be connected to a nearby ground.

If your car uses an alternator and there are clicking noises while accelerating, have the alternator and regulator checked in a competent auto service shop. Adding capacitors as described above should be avoided as they can damage the alternator.

FIG. 10-3 Filter for voltage regulator. (Courtesy Gold Line Connector, Inc.)

Antennas

FIG. 10-4 Feed through noise suppressor filter. (Courtesy Gold Line Connector, Inc.)

The filter in Fig. 10-3 can be used to minimize noise from ignition coils as well as accessories such as directional signals, blowers, and windshield wiper motors. Simply connect the filter between the hot lead at the accessory and a nearby ground. The feed-through filter in Fig. 10-4 can perform the function more efficiently than an ordinary capacitor, which has only one lead connected to each plate. It should be noted that accessories with motors generate whining or buzzing noises in the transceiver rather than clicking sounds. These noises are not affected by the speed of the engine.

Whining noise can be attributed to the generator or alternator if its frequency varies with the speed of the engine. This interference is usually minimized by cleaning or replacing the worn brushes. Filter capacitor GLC-1044 in Fig. 10-5 is designed to help eliminate this type of noise. Connect it between the output terminal on the alternator and a nearby ground.

FIG. 10-5 Whining noise suppressor filter for alternator. (Courtesy Gold Line Connector, Inc.)

Antennas

Once the noise problem has been eliminated, check the performance of the transceiver. If reception is poor or the transmitting range is inadequate, the problem is usually in the antenna system. Although the prime

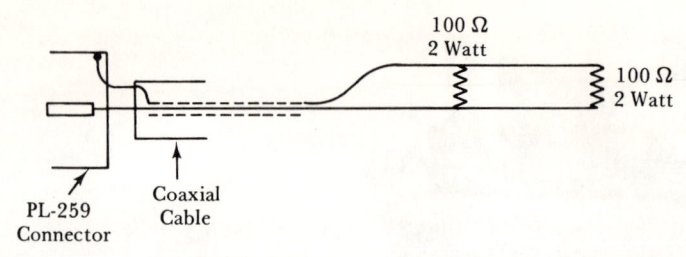

FIG. 10-6 Check coaxial cable using 50-ohm load to represent the antenna. Use two 100-ohm, 2-watt carbon composition resistors in parallel to form the 50-ohm, 4-watt resistor. Only noninductive resistors such as the carbon composition type should be used.

cause of a poorly performing installation is an inefficient antenna system, do not overlook inefficiency due to insufficient modulating voltage from a defective microphone. This may be the limiting item when you are not able to really "get out."

Antennas have been discussed in depth in Chapter 9. SWR should be adjusted before checking further for defects. If the SWR is not as low as you feel it should be, check the continuity of the antenna system with an ohmmeter. There should be about 0-Ω resistance between the center conductor of the coaxial cable and the antenna whip. (*Note:* Some antennas have a fiberglass outer shell or rod with a helical wound loading coil inside acting as the radiator. Fiberglass is an insulator. The ohmmeter reading for resistance between the coaxial and the outer rod of this type of antenna is infinite ohms.) A 0-Ω resistance reading is also required between the shield braid of the coaxial cable and the vehicle chassis. Repair all connections to the antenna and vehicle if a discrepancy exists in these readings.

When loading coils are used, zero resistance frequently exists between the inner conductor and the shield of the coaxial cable. You cannot know if there is a short in the cable unless you disconnect one lead of the cable from the antenna before checking resistance. It should, of course, be infinite.

You can make more checks on the cable only after it has been disconnected from the antenna. Connect a 50-Ω, 4-W carbon composition resistor at the antenna end of the cable and measure SWR in the usual fashion. If the cable is good, SWR is close to 1:1. See Fig. 10-6 for this setup. Here two 100-Ω, 2-W carbon composition resistors are shown connected in parallel to form the required 50-Ω 4-W load. This expediency is often necessary as 4-W carbon composition resistors are not readily available.

You will find that most problems with the antenna system are due to a defective cable or a poor ground. It is seldom the antenna itself or the loading coil that turns defective, although the antenna can rust and deteriorate. Just be sure that all connections and the installation have been made properly.

Power Supply Problems

It is easy to determine if power fails to reach the transceiver. Bulbs do not light or the pointer of the meter on the front panel of the transceiver does not deflect. If you are working on one of the few CB radios that has neither a bulb nor a meter, use a VOM to check for voltage at the power leads to the transceivers. Lack of voltage is usually due to a defective fuse. Replace it with a similarly rated fuse to properly protect the equipment. Be sure to use an instantaneous or slow-blow fuse, as required. If the fuse is intact, you must trace the wiring from the battery to the transceiver to determine where the break exists.

The amount of voltage supplied to the transceiver is important. If all is working well, voltage from the regulator should not exceed 14.4 V when the engine in the vehicle is running rapidly, nor should it drop below 12.2 V when the engine is idling. Some transceivers become unstable if the supply exceeds 14.4 V. Should the regulator need adjustment or the generator or alternator need servicing, it is recommended that to do the job properly the vehicle be taken to a competent auto service station rather than to an electronics man.

Ordinary lead–acid storage batteries can be checked using a hydrometer. The liquid or electrolytic in each cell has a *specific gravity*. It is this factor that is determined from hydrometer readings. Check each cell individually to be sure that they are all in good shape. The specific gravity of each cell depends on how well it is charged, as shown in Table 10-1. If the reading on one cell differs radically from readings on all other cells, or the specific gravity remains low after recharging the battery, it is defective and should be replaced.

Table 10-1 Percentage of Charge in Lead–Acid Storage Cell as Related to the Specific Gravity and Voltage of Each Cell

Specific Gravity Reading	% of Maximum Charge	Voltage Across Each Cell	Voltage Across 12-V Battery
More than 1.26	100	2.10	12.60
1.225 to 1.26	75	2.07	12.42
1.19 to 1.225	50	2.03	12.18
1.155 to 1.19	25	2.00	12.00
Less than 1.155	0	1.95	11.70

Index

AGC, 20, 132, 154
Alignment, 145-47
Alphabet, phonetic, 8
A-M, 14, 19, 20, 26, 88, 137
Antenna, 14, 27, 31, 32, 87, 185-207
  amplifier, 204-05
  base loaded, 200
  base station, 195-99
  center loaded, 200
  cophased, 32, 201, 206
  cophaser, 40
  directional, 185, 188, 189
  disappearing, 202, 203
  diversity reception, 32

Antenna *(cont.)*
  efficiency, 40, 190, 200
  feed point impedance, 186, 195, 196
  gain, 190, 191, 197, 202, 203
  ground plane, 189, 196, 203
  gutter mount, 31
  half wave, 31, 186-88
  impedance, 31, 186
  jack, 138
  loading, 25, 31
  loading coil, 31, 200, 201, 216
  loading, improper, 25
  magnet mount, 203, 210
  matcher, 38, 207

Antenna *(cont.)*
   mismatch warning lights, 25
   mobile, 200-05
   phased array, 198
   polarized, 197
   quad, 198
   quarter wave, 31, 201
   radiation angle, 189, 198, 199
   selector switch box, 32, 39
   troubleshooting the, 215-17
   tuning the, 205-07, 210
   yagi, 197
Audio
   amplifier, 88, 114, 122, 123, 143-45, 154
   modulation limiter, 118, 119
   modulator, 14, 114-19
   SSB, 122, 123
Automatic Noise Limiter, 21, 138, 139, 210
AVC, 19-21

Base station radios, 44, 45
Beta, 58, 59
Boom, 47, 175
Bracket, mounting, 211, 212
Buffer amplifier, 88, 110, 154
Bulb indicators, 155, 162

Cable
   coaxial, 185, 192, 193, 195, 197, 206, 216
   connector, 193-95
Calling channels, 4
Capacitor, coupling, 162
CB jargon, 8-12
CB license classes
   Class A, 1, 2
   Class B, 1-3
   Class C, 3
   Class D, 1, 3
   Class E, 3

Channel frequencies, 4
Codes, 7
Conversion, dual, 22, 94, 129, 130, 151, 154
Counter, digital, 77, 79, 80
Crystals, 101, 106, 107, 156, 163

Definitions, CBer's language, 8-12
Delta tune, 22, 24, 135-37
Detector, 137, 138, 140, 141, 154
Detector, product, 149, 150
Detector, troubleshooting, 140, 141
Digital circuit, 97
Diode, junction, 56, 57, 131-33, 138, 140
Dip meter, 81, 82
Dipole, antenna, 186, 187
Distortion, 116, 117
Drift, 5
Driver, 88, 154

Electronic System Noise, 212-15
   clicking, 213-15
   popping noise, 213
   whining noise, 215
Emergency channel, 2, 3, 9, 24

FCC rules, 3, 5, 6, 30, 87, 185, 191
FCC channel assignments, 4
Field strength, 36
Field strength meter, 34, 36
Filters
   TVI, 43
   high pass, 44
   crystal, 124-26
Frequency
   accuracy, 79, 80, 156
   CB channels, 4, 79
   counter, digital, 77, 79-81
   synthesizer, 89-95, 108, 147, 149

Index

Generators
  audio, 78, 79
  RF, 76-78, 105, 156
Ground plane, 31

Harmonics, 92, 94
Headset, 46, 47, 175-77
HELP, 2

I-f, 91, 134, 135, 155
Impedance, 186
Installation
  permanent, 211, 212
  temporary, 210, 211
Interference, 5, 6, 212-15

Jargon, CB, 8-12

License, 4, 5
Lightning arrestor, 41, 42
Load, dummy, 37, 83, 86
Loudspeaker, 26
  jack, 26, 46
  remote, 26, 46, 47, 166

Maritime channels, 3
Microbars, 28, 166
Microphone, 27-30, 88, 169-84
  base station, 173, 174
  ceramic, 27, 28, 169, 173
  compressor amplifier, 180, 181
  connector wiring, 170-73
  crystal, 27, 28
  dynamic, 27, 28, 169, 173
  hangar, 212
  noise cancelling, 174-76, 179
  oscillation problems, 179
  peak redistribution modulation, PRM, 181-84
  power, 30, 78, 177
  specifications, 28

Microphone *(cont.)*
  switching circuit, 14, 16, 105, 170, 171
  transceiver controls, all on Mic, 166, 167
  troubleshooting, 170-73
mixer, 91, 107-10, 134, 151, 154
Mobile CB radio, 45
Modulation, 14
  amplifier, 18
  amplitude, 88
  automatic limiter, 117-19, 165
  distorted, 165
  light indicators, 25
  limiters, 6, 30, 118, 119
  meter, 42, 43, 80
  microphone requirements for, 28, 30
  poor, 117, 165
  trapezoidal pattern, 72, 73
Monitor, automatic channel, 9, 24
Multimeters
  DVOM, 50
  TVOM, 50-53, 110
  VOM, 50-52, 86
  VTVM, 50, 52

Noise amplifier for squelch, 21
Noise, automotive
  clicking when accelerating, 213, 214
  popping, 213
  whining, 215
Noise blanker, 139, 140
Noise clipper, 21

Ohm's Law, 186
Oscillator, 88
  adjustments and troubleshooting, 102-07
  colpitts, 102, 103
  crystal, 87-89, 92-107, 119
  frequency synthesizers, 89-97, 134, 135, 147, 149

Oscillator *(cont.)*
   local, 134, 154
   phase locked loop, 89,
      97-100, 119
   Pierce, 102, 103
Oscilloscope, 70-74, 116
Overmodulation, 42
Overtones, 92, 94, 103

PA, 26, 78, 166
Power circuits
   AC supply, 13, 45, 158-61
   battery, 13, 45, 217
   troubleshooting, 217
Power, dc calculations, 186
Power, rf, 34
   amplifier, 87
   bulb indicator, 37, 38, 112
   meters, 35-37, 112
Power supply, 13
   Ac to dc, 45, 46, 158-63
   filters, 158, 160, 161
   full wave, 159
   half wave, 158
   regulated, 53-55, 160
   ripple, 13, 45
   voltages, 54
Push-pull, 114

Radiation, polarized, 187
Receiver, 14, 129, 130
   alignment, 145-47
   audio section, 14
   dead, 162
   detector, 14
   distorted, 163
   dual conversion, 22, 129, 130,
      147, 151-54
   hum, 158-61
   I-f amplifier, 14, 20, 134,
      135, 154
   intermittent, 161, 162
   local oscillator, 14, 19

Receiver *(cont.)*
   noise, 131, 163
   poor selectivity, 163
   Rf amplifier, 14, 130, 131, 147
   Rf gain control, 22
   single conversion, 129, 130
   SSB 147-50
   switch, local/distance, 134
   troubleshooting, 156-63
   weak, 162, 163
Resistance, 186
Rf, 87, 88
   amplifier, 88, 110, 111, 130-33
   gain control, 22, 134
   output stages, 154
   transformer, 133
   troubleshooting, 133
Rotator, antenna, 185, 192, 197

Semiconductor, 56
Signal injector, 75, 76
Signal tracer, 74, 75, 116, 117
Single sideband, SSB, 26, 89,
      119-28, 147-50
   ALC, 127
   amplifier, linear, 126, 127
   balanced modulator, 121, 123
   channels, 3
   crystal filter, 121, 149
   detector, 149, 151
   exciter oscillator, 121, 122, 149
   frequency synthesizer, 147, 149
   mixer, 126
   oscillator, 121
   PEP, 127, 128
S-meter, 24, 25, 155
squelch, 20, 21, 141-44, 154, 155,
      163
Specific gravity, 217
S/RF meter, 25, 155
Structures, antenna mounting,
      185, 191, 192
Switch, channel selector, 106, 107

Index

Switching circuits, 14, 90-96, 114
   electronic, 16-18, 154
   relay, 16
SWR
   description, 25, 31, 34
   meter, 25, 34-37, 80, 86
   tuning for minimum, 205-07, 216

Tank circuit, 97, 98, 101, 103
Telephone handset, 177
Temperature extremes, effects of, 209
Ten code, 7
Test instruments, 49-86
   bench, 50
   field, 50, 85, 86
Tone control, 26, 144
Transceivers, antenna coupler to, 39, 40
Transformer, audio, 154, 162
Transistor, bipolar, 58-63, 108, 110
   bias, 61, 62
   common base, 61
   common collector, 61
   common emitter, 61

Transistor, bipolar *(cont.)*
   Npn and Pnp, 58
   tester, 55, 62-65
Transistor, FET, 65-68, 108, 132
   IGFET, 65-68, 132
   JFET, 65-68
   tester, 68, 69
Transmission line, 34
Transmitter, 14, 15, 17, 18, 87
   A-m, 88
   low output, 164, 165
   no output, 164
   Rf amplifier, 14
   Rf oscillator, 14
   servicing restrictions, 5, 87
   SSB, 119-28
   troubleshooting, 163-65
   unstable with power mic, 179
Tube tester, 69, 70
Tuned circuit, 131-33
Tuning, 20
Type acceptance, 5

VCO, 97

Wattmeter, rf, 82-85
Wavelength, 186, 187